The Food Consumer

The Food Consumer

Edited by
CHRISTOPHER RITSON
LESLIE GOFTON
and
JOHN McKENZIE
Department of Agricultural and Food Marketing
University of Newcastle upon Tyne

A Wiley-Interscience Publication

JOHN WILEY & SONS
Chichester · New York · Brisbane · Toronto · Singapore

Library of Congress Cataloging-in-Publication Data:

The Food consumer
'A Wiley-Interscience publication'
Includes index.
1. Food. 2. Food habits. 3. Food industry and trade
I. Ritson, Christopher.
TX353.F597 1986 641 86-4092

ISBN 0 471 90984 X

British Library Cataloguing in Publication Data:

The Food consumer.
1. Food consumption
I. Ritson, Christopher
339.4'8'6413 HD9000.5

ISBN 0 471 90984 X

Printed and bound in Great Britain

Contents

PART III PRESENTING FOOD

List of Contributors

Christopher Ritson is Professor of Agricultural Marketing, and Head of the Department of Agricultural and Food Marketing, University of Newcastle upon Tyne, England.

Leslie Gofton is a Lecturer in the Department of Agricultural and Food Marketing, University of Newcastle upon Tyne.

John McKenzie is Rector, London Institute, and Visiting Professor, Department of Agricultural and Food Marketing, University of Newcastle upon Tyne.

Tim Josling is Professor, Food Research Institute, Stanford University, United States of America.

David Blandford is Associate Professor, Department of Agricultural Economics, Cornell University, United States of America.

Arnold Bender is Emeritus Professor of Nutrition, University of London, England.

Stefan Tangermann is Professor fur Agrapolitik, Institut fur Agrarokonomie der Universitat Gottingen, Federal Republic of Germany.

Paul Rozin is Professor, Department of Psychology, University of Pennsylvania, United States of America.

April Fallon is Assistant Professor, Department of Psychiatry, The Medical College of Pennsylvania, United States of America.

Marcia Pelchat is Assistant Professor, Department of Psychology, Washington College, Maryland, United States of America.

Anne Murcott is a Lecturer in the Department of Sociology at the University College, Cardiff, and in the Department of Psychological Medicine, University of Wales College of Medicine.

The late **David Lesser** was a Lecturer in the Department of Agricultural and Food Marketing, University of Newcastle upon Tyne.

David Hughes is Senior Agri-business Adviser, Development Alternatives Inc., Washington D.C., United States of America.

David Marshall is a Research Associate in the Department of Agricultural and Food Marketing, University of Newcastle upon Tyne.

Ray Bralsford is Technical Development Director, Cadbury Limited, Birmingham, England.

Elizabeth Nelson is Chairman, Taylor Nelson Applied Futures Limited, Epsom, England.

George Glew is Professor and Head of the Department of Catering Studies, and Director of the Hotel and Catering Research Centre, Huddersfield Polytechnic, England.

Judie Lannon is Research Director, J. Walter Thompson Company Ltd, London, England.

Preface

There are many definitions of marketing. A paraphrase of typical definitions would read something like 'the process whereby an organization, in order to fulfil its objectives, accurately identifies and meets the requirements of its customers'. Thus, central to the subject of food marketing must be the study of the behaviour and requirements of the consumer of food, the ways in which organizations attempt to meet these consumer requirements, and the impact of public policy on the consumption of food.

This book arose out of a realization within the Department of Agricultural and Food Marketing at Newcastle University—stimulated by the Life Sciences Division of John Wiley—that there was no volume which provided an integrated approach to the study of food consumption in the industrial world. The original approach from the publisher envisaged that we should ourselves write such a volume; our view was that it would be more effective to invite specialists to write individual chapters conveying the contribution of their own subject to the study of the food consumer.

Thus the book has two objectives. First, it aims to present, in one volume, a total picture of the consumer and his food in the Western world. Second, the intention is that it will be a book to which specialists will turn when looking for an introduction to an aspect of food studies which lies outside their own particular specialization.

The book is divided into three parts. As a foundation, the first three chapters introduce three broad areas, sometimes touching on issues covered subsequently in more depth. The first chapter considers the impact of public policy on food consumption. The second provides an overview of the major foods consumed in the industrial countries and the way broad patterns of food consumption are changing. Part I concludes with an introduction to the relationship between the foods eaten and the nutritional welfare of people.

Part II attempts to convey the contribution of four social sciences—economics, psychology, anthropology and sociology—to an understanding of the behaviour of the food consumer. Typically, these subjects have pursued their own line of inquiry, ignoring the others. This is particularly so when one considers the contribution of economics, as opposed to the other three subjects, which are sometimes known in marketing as the 'behavioural sciences'. For this reason Part II concludes with a chapter which attempts to provide an integrated view of the determinants of food choice, cutting across the traditional disciplinary boundaries.

In Part III we look at various aspects of the way organizations do attempt to identify and meet consumer requirements in food products.

Christopher Ritson
University of Newcastle upon Tyne
June 1985

Acknowledgement

The editors wish to record their appreciation of the contribution made to this book by David Lesser. His intellect, imagination and energy are deeply missed by his colleagues in the Department.

PART I
Food and People

The Food Consumer
Edited by C. Ritson, L. Gofton and J. McKenzie

CHAPTER 1

Food and the Nation

TIM JOSLING and CHRISTOPHER RITSON

INTRODUCTION

Public policy towards the food consumer reflects the political philosophy of a country, tempered by the specific concerns of the electorate. In Western democracies, the ethic of individual freedom of choice has ensured only minimal government intervention in food purchase and consumption decisions. The food industry is squarely within the private sector. Even the farming sector, laden with government programmes to influence output and incomes, is made up of numerous small businesses outside government control. Only in times of war have governments assumed a measure of direct involvement in the production and purchase decisions relating to food.

This view—that decisions about food production, distribution and consumption should remain outside the sphere of government activity—accounts, in part, for a seeming ambivalence in policy towards food. Clearly, what people eat is of vital importance. The nutrition of a population must rank, along with education, housing and defence, as a primary concern of legislation. Yet, while health, education, social services and, often, housing have been absorbed into the public sector, the provision of food has not. Most countries claim to have 'food policies', but upon examination these policies rarely involve more than the supervision of private sector activities. To the compulsive interventionist, this may appear a failure of collective action; most people in the mainstream of political thought, however, presumably see it as reflecting broad satisfaction with the ability of the private sector to meet the most essential of human needs, and the conviction that conflicts between practice within the food sector, and the public good, can be resolved by supervisory legislation.

The absence of an interventionist food policy in most Western nations does not, however, mean that government actions do not have an impact on the food sector and family consumption decisions. Three aspects of government influence on food consumption decisions can be isolated for particular discussion in this chapter. The first is the impact of policy on the *availability of food supplies*. Clearly, any action which affects the supply of domestic farm products, or the access to foreign products, has a potential impact on consumers. Such actions are commonly associated with agricultural policies. Their effect on food consumption is largely incidental. The second aspect is the attempt by governments to alter the *purchasing power of food consumers*. Governments take an interest in the price of food and in the incomes of groups within the economy that might have difficulty in maintaining an adequate diet. The third aspect of government policy discussed in this chapter is the *control of food quality*. This control manifests itself both as an attempt to mandate quality through standards and health safeguards, and in the provision of information on nutrition to enable individuals to take health aspects into account when making food consumption decisions. The extent to which consumers as such play a role in influencing decisions on these three aspects of food policy is touched upon in the last section of the chapter.

POLICIES RELATING TO FOOD AVAILABILITY

The typical supermarket in an affluent economy contains aisle upon aisle of foodstuffs in a great variety of forms and from many origins. The manager of the supermarket has an incentive to keep the shelves well stocked, and to maintain the profusion of offerings. The government also has an interest in ensuring the ready availability of at least the basic foodstuffs, but in general it takes this for granted. Though governments do manipulate food markets they mostly do so for other purposes. The extent to which government actions help or hinder the work of the supermarket manager is the topic of this section. The effect depends largely on the type of product and the raw material from which the food product is made.

Consider first the fresh produce section of the supermarket. This will typically contain seasonal fruits and vegetables from local farms as well as produce from abroad, often of a type that cannot be produced locally. The availability of these two groups of commodities is more likely to be constrained by the weather than by government action. European governments do not in general put barriers in the way of imports of 'non-competitive' tropical produce, and may even welcome such trade as assisting the exports of former colonies and other developing countries. Local produce, too, is likely to have arrived on the shelf with minimal government involvement. However, at times when domestic supplies are heavy, many governments have policies designed to assist producers to remove some of the crop from the market. Though such policies occasionally make the headlines, the concept of denying to the consumer the benefit of a bumper harvest is accepted with surprisingly little question.

The most significant policies affecting the availability of fresh produce relate, not to clearly local or unambiguously foreign goods, but to the area of overlap between these categories. Governments, through border controls, commonly reduce the supply of competing foreign fruits and vegetables in order to increase the demand for local products. This often takes the form of reserving a 'season' for the domestic goods. By extending this 'season', local producers can get higher prices for 'early' produce. Protection of local producers can also take the form of import quotas imposed on various commodities, and of 'reference prices' which allow for taxes on imports if the produce is offered at a price considered too low by those concerned with local farm incomes (Ritson and Swinbank, 1984). All these measures reduce the availability of foreign goods which the consumer might prefer, or raise the price of local goods by exposing them to competition. International specialization of production in fruits and vegetables, to take advantage of climatic differences, has led to a situation where many kinds of produce are available for much of the year. If the supermarket manager can only provide, say, strawberries for his customers for a few weeks in summer, the chances are that the hand of the government is at work. Though diets are not threatened, there is a significant loss of choice and of potential consumer satisfaction from such protectionist policies.

The meat counter in the typical supermarket is also likely to reflect government policies towards the producer. Most countries raise the bulk of the beef, pork and chicken that are sold to domestic consumers, though sheepmeat production is somewhat more specialized. Agricultural policies increase the demand for the local product by taxing imported goods, or by entering into agreements with foreign countries to limit such imports. Domestic production of meat can vary over time, as an effect not so much of the weather as of the stage of the product 'cycle'—the mysteriously coordinated actions of producers building up or running down their herds or flocks at the same time. Imports provide a valuable mechanism for stabilizing supply, and this is on occasion reinforced by government policy. The European Community (EC), for instance, reduces the tax on imported beef when domestic prices are high to encourage supplies from abroad.[1] In some countries, subsidies for domestic storage of meat also help to even out the variations in domestic production.

Except in times of serious economic disruption, the availability of meat on the supermarket shelves is unlikely to be a major issue. Consumers switch with relative ease between, say, beef and pork, or between chicken and red meat. The impact of government policy is mainly to make all meat more expensive, as a consequence of restrictions on trade. However, the main beneficiaries of this protection are not always the producers of meat. Animals are 'consumers' of agricultural products, and a large part of the income from selling the livestock products goes to pay for animal feed. In areas where feed costs are kept high by government policy (the EC is a notable example) the same policy commonly uses restrictions on meat imports to allow livestock producers to recover these extra feed costs from the consumer.

This takes us to the cereal products section of the store. Grain has always played a central part in government food policy. International shipments of wheat, maize, rice and other grains are the most convenient way of supplementing the food production capacity of countries not blessed with the resources to be self-sufficient. Unfortunately, the regulation of importation of grain offers an easy way to boost the incomes of local grain producers: apply tariffs or levies to such imports and the food industry will pay the higher price for domestic as well as for foreign grain. These higher prices show up in more expensive bread, cakes, breakfast cereals and biscuits for the consumer, in addition to the higher priced pigmeat, poultry and eggs. Exporting countries do not have the easy option of limiting imports, and tend to have a lower structure of grain prices.

The importance of trade in grains—desirable in itself as a way of offsetting both year-to-year fluctuations in local output and inherent differences in agricultural capacity—has brought with it perceived problems. These relate first to the ability of countries to pay for these imports, and second to the reliability of such flows in the future. In the case of the first concern, some real problems exist for poorer countries, where the capacity to pay for any imports is limited and where international market fluctuations might have a disproportionate impact on their ability to maintain food supplies. (See, for example, Josling, 1980; Valdes, 1981, and Koester, 1982.) This interpretation of 'food security' is a much less immediate problem for consumers in developed countries. Despite some success by the farming industry in promoting 'import-saving' arguments for protection, it is now generally recognized that payment imbalances are caused by particular monetary and exchange rate policies, and their cure should similarly be economy-wide in scope. Thus, it can be concluded that consumers should be asked to pay the 'full' price for imports of grain (i.e. at the exchange rate necessary for international payments to be balanced); however, to impose prices in excess of that is likely not only to decrease national income, but also, paradoxically, sometimes to weaken the overall payments balance. (See, for example, Josling, 1970, and Phillips and Ritson, 1970.)

To answer the question whether supplies on international grain markets are reliable requires clairvoyance rather than economic theory. During the 1960s, the international price of grain stayed remarkably constant, despite monsoon failures in Asia and other harvest fluctuations. The US retained significant surplus capacity and more than adequate levels of stocks, while in Europe the output of cereals increased quite rapidly. This situation changed dramatically in the early 1970s, when a combination of high demand, Soviet purchases, poor crops and economic uncertainties conspired to send grain prices up sharply. With the exception of one or two years, the relative price of grain has fallen since that time, and the situation of 'surplus' in grain markets has reappeared. Clearly, anyone who considers that another, more permanent, 'world food crisis' is just around the corner could argue for a high level of domestic production in anticipation of that event. Since the odds appear to be against the re-emergence of shortages in the near future, the costs of guarding against this eventuality are high, relative to the benefits. 'Security of supplies' may be a useful phrase to bolster the case for farm income support, but as an element in a consumer-oriented food policy it has doubtful merit.[2]

Consider next the dairy case: a mixture of fresh produce (milk) and processed goods (butter, cheese and yoghurt) is of considerable importance to the consumer. In Western countries it is for these products that the hand of government is most clearly visible, aiming not at ensuring supplies at reasonable prices, but at deriving the maximum revenue with which to reward the producer. Dairy output has, in the past, often come from small farms in regions not well suited to the production of cereals. The daily sale of milk has represented an essential, if modest, source of income to these farm families. For better or for worse, the situation has changed in recent years. Though small farmers are still numerous, there are fewer of them, and medium- and large-scale enterprises have sprung up to provide the bulk of milk produced in industrialized countries. Furthermore, the intensity of production has increased as a result of a shift from pasture to compound feed as the basis for milk output. Such progress is not necessarily unwelcome in itself, but ironically it has been promoted by price support policies ostensibly geared to the small-farm income problem.

Two basic market management mechanisms are in use in connection with dairy policies in developed countries. One is to make use of the relatively inflexible demand for liquid milk—due largely to the absence of any close substitute product—to raise milk prices to the consumer and increase revenue from such sales.[3] This can be done most readily by the use of supply control through marketing boards, licensing systems or fluid milk contracts. The milk 'surplus' to liquid requirements is then sold to processors at a lower price.

The other mechanism is to support the market for milk products (with no differentiation in price between this milk and that for liquid consumption) by control of imports or by subsidies for exports or (temporarily) government-financed storage.

The first mechanism is used where markets are reasonably well-organized, say through cooperatives or other institutions with exclusive milk collection rights; the second is found where the liquid market is not so strong or where organization of producers has not proved possible. Both exploit the consumer, as purchaser either of milk or of milk products: but neither system has been able to cope with the flood of milk from the 'modern' dairy sector.

In contrast to the growing international market for grains, the outlets for surplus dairy produce are strictly limited. Much of the world either for physiological reasons cannot drink milk, or would have no way of keeping this perishable product fresh. Technology can overcome this second problem, but the prospects for lucrative export markets are dim. Hence, considerable subsidies are needed to remove from the market the produce that domestic consumers cannot afford to buy. It is ironic that in many cases it is cheaper, in terms of government programme costs, to dump surpluses on an unwilling foreign market than to release the same products onto the domestic market. Consequently, the 'world market' for dairy products, unlike that for most agricultural products, is conditioned essentially by the willingness of a few countries (such as the USSR) to soak up the surpluses of others, at 'bargain basement' prices. There are no problems of availability of milk products to industrial country consumers, only problems arising from too much milk relative to the willing commercial demand at the high prices set by policy.

Other supermarket shelves show less evidence of government policy, either for or against the interest of the food consumer. Farm product cost is generally a low proportion of retail price for frozen, canned or otherwise packaged goods and for many convenience and highly processed foods. As a consequence, these items tend to be less involved in the policy mechanism for passing on the costs of farm support to the retail level. In addition, continuity of supply is both less of an issue for the country and less of a problem for the food industry. A shortage of baked beans, cake mixes or TV dinners is hardly likely to threaten the health of the country and the relative ease of storage and transportation of such processed items makes such shortages rare.

One final category of goods does, however, bear comment. The tropical beverages (tea, cocoa and coffee), together with sugar, share the characteristic of sharp fluctuations in availability and price on world markets. In the case of tropical beverages there have been various attempts at operating international commodity agreements to help stabilize prices. Under the auspices of these agreements, exporting countries accept limitations on exports at times of depressed prices. Importing countries have generally participated in these pacts, more from foreign policy motives than because they hold out any promise of alleviating any future supply shortages.

Sugar, as a commodity, spans both tropical and temperate zone farming. To the consumer, there is no difference between tropical (cane) sugar and that refined from sugar-beet grown in temperate climates. Farm policy, however, has been steadfastly geared to maintaining domestic production in northern countries, encouraged by successful lobbying on behalf of a relatively small number of producers and spurred on by fear of world shortages. Against this pressure, overseas development policy considerations have argued for continued markets for tropical sugar. The food consumer, while benefiting in the occasional shortage years, pays the bill for both the generous level of domestic income support and for the overseas aid component of high prices to developing countries. As with grain, a degree of price stability and supply assurance for consumers is obtained at considerable cost in terms of prices higher than necessary to induce adequate supply.

For most basic foodstuffs, then, the best way of assuring ready availability to the consumer is to encourage the harmonious development of international trade in those products. Many of the price-raising measures discussed above do introduce an element of stability in consumer prices. Unfortunately, the same domestic measures that protect consumers from price changes inevitably put more strain on world market prices. World price stability is determined by the aggregate response of stock levels and consumption to changes in production. If domestic prices do not reflect scarcity and abundance, then neither will stocks and consumption. Those parts of the global market that are less 'isolated' must, perforce, respond more, since all production (unless destroyed) must end up either in storage or in consumption, in turn leading to further demands for isolation. This cycle needs to be reversed if trade markets are to retain their stabilizing function. To date, the degree of cooperation among even the most friendly of countries has not been adequate to start this process.

POLICIES RELATING TO FOOD PRICES AND CONSUMER EXPENDITURE

One stated objective of most industrial countries in formulating agricultural policy is 'to ensure reasonable prices in supplies to consumers'.[4] Yet it is generally accepted that the consumer influence in determining government policies towards food prices in developed countries is weak, relative to the influence of farmers' interests. At best, consumer concerns may moderate the extent to which governments would otherwise pursue price-raising policies. Only rarely do Western governments adopt policies specifically directed towards lowering (or restraining increases in) food prices.

The relatively low influence of consumer interests on government policies towards food prices may in part be due to the difficulty of establishing, both in the minds of consumers and in government thinking, what is 'a reasonable level of prices' for food. Clearly, it is much more feasible for farmers to identify (and for governments to recognize) a level of product prices necessary to generate an acceptable level of farm incomes, given certain farm costs. At some point food prices would be so high that malnutrition (in the sense of an inadequate food intake) would threaten a substantial proportion of the population. Such a situation is only too real in many low-income countries. But in the developed world, although it is still possible to find deprived groups which are unable to afford what nutritionists would regard as an adequate diet, food prices do not generally threaten nutritional health. Over-eating, and an inappropriate balance of food, are now accepted as more significant nutritional problems in most advanced countries. Paradoxically, therefore, high food prices may promote, rather than hinder, what nutritionists would regard as a healthy diet, at least for the majority of people living in developed countries.[5] It is probably true to say, therefore, that in developed countries government policies which influence food prices are not primarily motivated by concern over the general level or balance of food intake. Rather, it is accepted that, where particular groups would otherwise be unable to afford a nutritionally adequate diet, income supplementation through welfare and social security programmes is the solution.

This is not to say that there are no policies which keep down the prices of particular foods, nor that governments do not on occasion give direct relief to certain food consumers. Two types of 'consumer subsidy' policy can be distinguished, depending upon whether a foodstuff is made generally available at a lower price than it costs the supplier, or is available at a reduced price only to a limited group of consumers. General (or non-targeted) food subsidies are typically aimed at increasing demand and preventing or alleviating the build-up of surpluses. The European Community has tried various schemes over the years for reducing butter stocks. Consumer 'aids' are paid to olive oil traders to make it possible to sell that product in competition with other oils. The problem with such surplus disposal programmes, from a government's viewpoint, is that they can become costly. Making the consumer pay the high producer price for such items conveniently hides the cost of farm income support: asking the taxpayer to bridge the gap between a lower consumer price and a higher producer price brings this cost out into the open—though it does allow the cost to be presented sometimes as a 'consumer' rather than as a 'producer' subsidy.

One way around the problem of the cost of subsidizing all consumers of a product is to limit the subsidy to particular groups. This allows an element of nutritional planning along with the desire to reduce surpluses. Among the groups often chosen are those with special nutritional needs—young children and nursing mothers—or with economic needs—those on welfare or on state pensions. School lunch programmes, and the provision of low-price milk for schoolchildren, have been common in industrial economies. The mechanisms used to achieve the objective of limiting the benefits of consumer subsidies to particular groups range from distribution by agencies catering for those groups to vouchers or coupons redeemable by eligible individuals. Low-income families in the US can occasionally benefit from the release of government-held stocks of dairy products. The European Community has run beef subsidy schemes for pensioners. The more exactly a deprived or non-consuming group can be targeted, the less expensive the subsidy scheme is likely to be, the more likely it is to be tied to nutritional objectives, and the more effective it will be in increasing consumption.

One notable exception to the generalization that industrial countries do not have extensive food subsidy programmes for low-income consumers is the food stamp programme in the US. Conceived originally in the 1930s, with the twin objectives of maintaining the purchasing power of poor consumers and disposing of surplus farm commodities, it was reintroduced in 1961 with the more direct objective of abolishing hunger in a generally affluent society. The programme has blossomed since the 1960s in large part as a political counterbalance to the farm support policies. Urban congressmen support farm income policies in return for rural support of food stamp expenditure. Total programme costs now exceed $10 billion annually and a family with a very low income can receive up to $50 per week in coupons to be spent on food.[6] Over one-tenth of the nation's consumers benefit from the programme. In spite of this apparently wide coverage, many poor families do not make use of the food stamps to which they are entitled, whilst families which would not be generally thought of as poor have been able to benefit. This has led to some tightening of the eligibility criteria in recent years. The programme is now generally considered to be embedded in the general range of welfare measures and is, in many ways, as much an income supplement as a specific food subsidy to those who receive its benefits.

Given, therefore, that nutritional goals do not generally determine what is a 'reasonable' level of consumer food prices, other criteria must apply. Economists regard the price at which competing supplies are available from international markets as the appropriate yardstick. In practice, however, 'world prices' seem to have very little impact on consumers' awareness of whether or not they are paying reasonable prices for their food. One exception is where there has been considerable publicity for subsidized exports of food. Perhaps the most famous example of this has been the periodic sales of very cheap EC butter to the Soviet Union. The publicity attached to such cut-price sales attracts much greater significance in terms of government policies towards domestic food prices than does publicity concerning taxes on cheaper competing imports. From a consumer viewpoint, the principle of obtaining domestically produced food supplies at similar prices to those at which produce is available from international markets is of equal importance to that of domestic production not being made available to foreign consumers at lower prices than those paid by home consumers. However, the consumer is generally ignorant of the level of prices on world markets, and no attempt is made by public agencies to dispel this ignorance. Consequently, all attention is focused on the actual price in the shops rather than the potentially lower price in the absence of import restrictions or export subsidies.

If consumers appear to have little say in defining 'reasonable' prices for foodstuffs, one issue which does lead to political concern, and sometimes to political action, is that of *rising* food prices. If food prices are going up, irrespective of whether they are 'high' or 'low' to start with, and irrespective of whether or not they are rising faster than other prices, then they tend to assume national importance. Because of the regularity of purchases, food price inflation is more readily recognized as such by consumers. This has led to the view that, in breaking a 'wage/price inflationary spiral', food prices are a critical variable. If food prices are controlled, most probably this will be part of a general counter-inflation policy. One therefore sees such policies emerging at times of rapid inflation. Such a period occurred in 1973-1975, when world market prices for a number of commodities rose sharply. The UK, for instance, instituted a consumer subsidy programme to keep down the price at the retail level of a number of key foodstuffs (Ritson, 1975). The US experimented with price controls at the same time, while discouraging exports of farm products. Such policies have a high budget cost, and are usually removed quickly once world market prices return to more normal levels.

POLICIES RELATING TO FOOD QUALITY

Government intervention in order to meet food quality objectives may be grouped into three kinds. These are related to:

(1) the performance of the market in meeting consumer requirements;
(2) information supplied to consumers;
(3) consumer safety.

Quality standards are measurable properties of food which differentiate them for consumers. Grading is the sorting of varied kinds or qualities of produce into similar categories. Government imposed (or, more usually, encouraged) standardized food grades can be seen as a substitute for commercial branding, where the structure of the market is fragmented and dominated by small firms. Thus, the first category above is merely the extension towards the food consumer of government intervention in agriculture motivated by the possibility of market failure. It has long been contended that various features of agricultural production, and the structure of agricultural product markets, may require government intervention if markets are to operate efficiently. One aspect of this concerns the ability of the food-marketing chain to transmit accurately to producers information relating to consumers' quality requirements. Governments have reacted, therefore, by promoting grading and classification schemes. For example, quality grades applying to fruit and vegetables attempt to impose, on a traditional and fragmented market, means by which producers can identify, and consumers recognize, variations in product quality. Carcase classification schemes aim to convert consumers' preferences for different qualities of meat into recognizable criteria for livestock producers.

The second kind of intervention refers to legislation designed to ensure that food consumers can make 'informed' choices in their purchases. Broadly speaking, such legislation usually covers two areas. One reserves certain terms for products which meet specific composition requirements. For example, in the UK, 'orange juice', 'orange nectare', 'orange drink', 'orange crush' and 'orange squash' are all descriptive terms reserved for different products in terms of fruit composition.[7] The second covers legislation directed towards ensuring that food products are not labelled (or advertised) so as to mislead regarding nutritional or dietary properties (e.g. with respect to 'slimming properties').

It is, however, food safety legislation which would usually be regarded as the most important, and would seem to be the most problematical, area of food quality legislation. The problem is one of converting a health hazard, which will typically be one of degree, into specific rules. Safety legislation covers two broad areas—hygiene and additives. The former is directed towards the prevention of 'food poisoning' and legislation may be couched in terms of either microbiological standards (the number of bacteria in a given quantity of food), or methods of food preparation and packaging and storage in order to minimize the possibility of contamination. (These issues are discussed in more detail in the chapters by Bender and Bralsford.)

The question of government control of the use of additives is complex. Safety evaluation is expensive and ambiguous. The problem is well described by Grose (1983):

> No substance is absolutely safe. What really matters is the fate of any substance in the human body.... The toxicologists have tried to calculate the effect on man by feeding additives and other substances to animals. The duration of these studies for one additive is three to four years, and the cost has risen to £300,000 and more. The number of food additives awaiting further toxicity testing is said to be in excess of 200.... Now the toxicologists are questioning whether information derived from animal studies has any relevance in assessing the hazard posed by chemicals in terms of human exposure. In future evaluation must, as far as possible, be based on a substance's metabolic fate in man. Presented with the mounting costs of toxicological testing, and its uncertainties, the scientists and the food industry are looking closely at the disciplines of evaluation of additives. In the USA there have been various proposals for methods of ranking levels of risk, as information is gathered about the metabolism of substances in humans. Although the Americans would clearly like to be able to quantify risks and benefits, they seem to have backed away from strict cost/benefit analysis with its difficult and repugnant requirement that a monetary figure be allocated to a risk to an individual human being.

Another problem with additives, as with nutrients, is discussed in chapter 3. It arises from the fact that people eat a varied diet. Even if it were possible to determine a 'safe level' in a *particular* food, the importance of that food relative to others, or in combination with others, that the individual eats is the critical factor in assessing the safety of the diet.

There is one further area of potential government intervention concerned with food quality and this relates to what might be termed 'the quality of diet'. Until recent times this has seemed a fairly straightforward issue. If people's diets were inadequate, it was because of either lack of money or lack of knowledge. There could be little argument over the merits of government price and welfare programmes designed to improve the capacity of the poor to feed their families, nor of policies directed towards nutritional education. More recently, however, the focus has shifted to concern over the health implications of food in general and certain broad categories of food in particular. The question of what is a healthy diet and how to evaluate the evidence relating to this is one to which we return in chapter 3 of this book. Our concern for the moment is the debate over the role of government (if any) in improving the diet of the population when malnutrition, in the sense of an inadequate food intake, is no longer regarded as the primary problem.

The view that there is a need for developed countries to adopt food and nutrition policies is not new; during the 1920s the Health Section of the League of Nations recommended that all countries develop national food policies. But the role of government in influencing the nutritional quality of diets remains a controversial issue. Broadly speaking, this controversy concerns two issues: the degree of proof required before a government should publicize the merits of reducing the consumption of certain foods, and the question of 'freedom of choice'.

With respect to the former issue, clearly, there may be powerful farming and food industry interests which would be adversely affected by a strong trend in demand away from food products which are deemed suspect on health grounds. As discussed in chapter 3, this particularly concerns products containing a high proportion of animal fats (particularly dairy products) and sugar. In general, a government-inspired movement away from consumption of such products is a more serious issue than the banning of specified additives: the latter can usually be replaced, the firms concerned merely incurring extra costs. But when it comes to altering the balance of a nation's diet, the prosperity of major sections of the food and farming industries may be irreversibly affected. Governments are therefore understandably reluctant to embrace wholeheartedly the dietary recommendations of medical committees, when doubt remains over the evidence in support of the health hazards of, for example, sugar and fat consumption; and they are sometimes accused of actively suppressing information which may be damaging to powerful food and farming interests. For example, there was enormous publicity in the UK when the contents of the NACNE (1983) report were leaked to the *Sunday Times* newspaper. The UK Government was accused of suppressing the report. However, the similar findings of another report (COMA, 1984) have been endorsed by the UK Government, and are leading to minor policy initiatives—e.g. in the requirement that manufactured food products must indicate fat content.

On the issue of freedom of choice, to some people, the very idea of a nutrition policy is totally unacceptable and tends to produce an emotional response—it is the government 'telling people what to eat', removing a 'fundamental freedom of a democratic society'. Yet some kinds of government involvement seem to be regarded as perfectly permissible. For example, individuals who would condemn any attempt by a government to influence patterns of food consumption on health grounds may be quite happy that certain food additives should be prohibited; and individuals who would deplore the use of prices to influence food consumption patterns may not be opposed to a government financed advertising campaign directed towards nutritional education.

Behind such apparent contradictions there lies what is, in principle, a logical distinction. This is based on the view that it is permissible for a government to devote funds to raising the nutritional knowledge of the population; and it is permissible for a government to influence directly what food can be consumed when it is unreasonable to expect consumers to be able to assimilate information concerning the health properties of their diets; but no action should be taken to prevent individuals from eating whatever they wish, when they are fully aware of the health implications.

The distinction is intellectually satisfying. Regrettably, in practice the distinction is less helpful. Aspects of food consumption which have health implications do not always fall neatly into those which consumers can, or cannot, be expected to know about; and all societies do, in certain circumstances, take action to prevent individuals from doing what they know to be harmful—an example being the legal enforcement of the wearing of car safety belts.

Of the countries of the West, the case for the development of a nutrition policy seems to have been most widely accepted in Norway. In 1975 a government White Paper was published outlining a proposal for a comprehensive national food and nutrition policy (Blythe, 1978). This covered a variety of issues relating to, for example, food quality, self-sufficiency and rural prosperity. What distinguishes the Norwegian initiative, however, is the inclusion of the specific aims of reducing fat and sugar consumption (and increasing the intake of foodstuffs with heavy starch concentrations) and, more particularly, that price policy should be used to implement these aims.

> To encourage the production and consumption of desirable foodstuffs, production subsidies will be used as much as possible to regulate the production of those foods that are supplied mainly through domestic production, and consumer subsidies will be used to make those products that are desirable, but whose price is fixed on the international market, competitively advantageous on the domestic market. (Ringen, 1977)

We have yet to see whether the aims of the Government of Norway will be sustained in price policy over a prolonged period, particularly if a conflict arises between nutritional objectives and farming interests.

The Norwegian approach is very much the exception. Most governments of developed countries, while content to manipulate consumer food prices freely in order to meet agricultural objectives, do not use pricing to meet nutritional objectives. What is not generally realized, however, is that (by accident) the pattern of prices which results from agricultural policy is not often in conflict with contemporary nutritional goals. In 1981 one of the authors was invited to contribute a paper at a conference on Commercial and Political Influences on Dental Health (Ritson, 1983). It was clear in discussion that, having read of EC food surpluses, many people in the dental profession were convinced that the EC's Common Agricultural Policy was somehow promoting the consumption of butter and sugar by the British people. In fact, the effect of the UK adopting the EC's CAP had been to reduce British sugar and butter consumption, perhaps by as much as 10 per cent in the case of sugar, and more with butter. Though these accidental 'by-products' of farm support policies can be in tune with nutritional goals, one should not equate these effects with a coherent nutritional policy. In general, food prices reflect the wholesale cost of food products as influenced by government policy toward the producers, whilst the safety net of social programmes aims to provide families with incomes adequate to buy food. Government policy towards consumers is largely in the area of information and in the proscription of harmful substances.

THE CONSUMER VOICE IN POLICY

The three preceding sections of this chapter have illustrated the way in which governments in industrial countries have responded to the issues of availability, affordability and quality of food. As with all public policy areas, the role of government is forged out of the pressures of conflicting interests. This section returns briefly to the question of whether the food consumer as such plays a full part in the process of political compromise.

Students of the political scene have pointed out that individuals tend to identify more strongly with their employment than with their consumption interests in matters of political debate. We consider ourselves primarily as car workers, bankers, farmers or teachers and devote less effort to guarding our interests as buyers of cars, banking services, food and education. Though this may be a reasonable distribution of the limited time that we wish to spend on 'political' matters, it can lead to the underrepresentation of such consumer interests. Perhaps partly reflecting this producer bias in political influence, information on (say) agriculture and food is distributed by those who have a professional or bureaucratic interest in food production. Whether the lack of information available to consumers (on, say, the amount of hidden tax on food purchases through restricted access to world markets or through domestic support buying) is a cause or a consequence of the lack of political interest is a moot point.

In addition to the concerns most people have as producers and as consumers of goods and services,they also have a clear interest in the direction of government policy in their role as taxpayers. Indeed, this interest seems to be more clearly articulated and defended than the consumer aspects of policy. Thus we can think of the programmes discussed above as being a reflection of the strengths of these three interests, as they come together, to develop and finance new programmes and consider the continuation of present policies. The seemingly weak position of food consumers in influencing policy is the consequence of the strength of producer and taxpayer influences. There is no doubt that many of the complex problems of the farm sector could be addressed at far less real cost to the economy by the use of direct payments and producer subsidies. A deficiency payment system for grains in the EC to replace the existing import levies and export subsidies would, for instance, immediately remove all the implicit taxes on livestock farmers and consumers of grain and livestock products. Direct payments to small milk producers would likewise go far towards allowing a substantial reduction in consumer prices for milk and dairy products in the EC. However, governments, civil servants and, apparently, the electorate are more concerned to keep visible, financial policy costs low than to introduce more efficient methods of supporting farm incomes. Taxpayer interests dominate that of the food consumer in the political process.

Observers have for several years predicted the growth in political awareness and influence of consumer groups. In both the US and Europe there has been some such development. In the US it has taken the form of public advocacy of consumer causes, usually directed at large corporations and centred on the courts rather than the legislature. Producer interests have, if anything, strengthened their position in Congress through the financial assistance given by Political Action Committees. Urban lawmakers have acquiesced in farm policy decisions largely as a result of the success of the linking of the food stamp and the farm support programmes.[8] In Europe, the consumer movement has tended to emphasize information provision and product testing: attempts to become a partner in the discussion on farm policy in the EC have met with little success. Consumer voices are tolerated and patronized by the agricultural policy establishment: the power of producer groups is respected and feared.

Perhaps the greatest enigma in the politics of food is the lack of overt influence of the food industry *per se*. Less than one-half of the consumer price of foods goes for the purchase of raw materials—the rest is absorbed in the processing and marketing sectors. Employment in these activities is overtaking that of the agricultural sector. So long as the additional cost of raw material can be passed on to the consumer, there may be little incentive to take the consumers' part in the policy conflict. But if ever the food industry decided to embrace wholeheartedly the interests of the food consumer as being consistent with its own, there could be a major change in the tenor of the farm and food policy debate.

NOTES

1. This is in addition to the reduction of the import tax when *world* prices are high, the normal EC method of preventing the transmission of world price variations onto the domestic market. For details, see Harris, et al., 1983, Chapter 5.
2. For a detailed discussion of self-sufficiency and food security, see Ritson, 1980.
3. If the demand were more flexible (i.e. elastic with respect to price), revenue could be increased by *lowering* the price and increasing sales by a more than proportionate amount. Inelastic markets are an invitation to restriction of supply: elastic markets tend to produce a more competitive response on the part of producers. The chapter by Tangermann in this volume considers the significance of demand elasticity more fully.
4. This particular phrase is taken from Article 39 of the Treaty of Rome, setting up the European Economic Community. The community's Common Agricultural Policy is considered to be one of the industrial countries' least solicitous agricultural policies towards the consumer, despite this objective.

5. This statement perhaps needs some qualification in that it refers to the price-raising impact of agricultural policies, affecting, in particular, dairy products, meat and sugar. What nutritionists would regard as 'health versions' of industrial products—lean meat, low-fat dairy products, wholegrain cereal products, sugar-free canned goods and so on—are often more expensive, but this is not usually because of agricultural policies.
6. Clearly, there is some benefit to farmers from the increased consumption, but the food stamp purchases are not restricted to surplus commodities or even to domestically produced foodstuffs. In that respect it is a genuine consumer policy rather than a disguised farm support policy. For more details of the food stamp programme, see Timmer and Nesheim, 1979, and Timmer et al., 1983.
7. This also provides an example of the frequent gap between policy objectives and achievements in this area. Few consumers are aware of the meaning of such distinctions. However, there is evidence that some improvement is on the way. Following the publication of the COMA (1984) report, the British Government is introducing legislation on food labelling, and is carrying out a considerable amount of market research to determine the best method of presenting the information.
8. This has enabled successive Food and Agriculture Acts to gain the support of both rural and urban interests. The reaction of the taxpayer, now that these programmes are becoming more expensive, is just beginning to be felt.

REFERENCES

Blythe, Colin (1978) 'Norwegian nutrition and food policy', *Food Policy,* **3** (3) August.

COMA (Committee on Medical Aspects of Food Policy) (1984) 'Diet and cardiovascular disease', Report on Health and Social Subjects no. 28. London, DHSS.

Grose, Daphne (1983) 'Food standards legislation as a policy instrument'. In *'The Food Industry: Economics and Politics',* ed. J. Burns, J. McInerney and A. Swinbank. Heinemann, London.

Harris, Simon, Swinbank, Alan, and Wilkinson, Guy (1983) *The Food and Farm Policies of the European Community.* Wiley, Chichester.

Josling, Timothy E. (1970) *'Agriculture and Britain's trade policy dilemma'.* Thames Essay 2. Trade Policy Research Centre, London.

Josling, Timothy E. (1980) *Developed-country agricultural policies and developing-country supplies: the case of wheat'.* Research Report No. 14. International Food Policy Research Institute, Washington, DC.

Koester, Ulrich (1982) *'Policy options for the grain economy of the European Economic Community: implications for developing countries'.* Research Report no. 35. International Food Policy Research Institute, Washington, DC.

NACNE (National Advisory Committee on Nutrition Education) (1983). Proposals for nutritional guidelines for health education. London Health Education Council.

Phillips, Truman, and Ritson, Christopher (1970) *'Agricultural expansion and the U.K. balance of payments', National Westminster Bank Quarterly Review,* February.

Ringen, K. P. (1977) *'The Norwegian food and nutrition policy',* American Journal of Public Health, **67** (6).

Ritson, Christopher (1975) *'Who gets a subsidy?' New Society,* **31** (642), January.

Ritson, Christopher (1980) *'Self-sufficiency and food security',* CAS Paper no. 8. University of Reading.

Ritson, Christopher (1983) 'The effect of agricultural policies', *Journal of Dentistry,* **11** (2).

Ritson, Christopher, and Swinbank, Alan (1984) 'Impact of reference prices on the marketing of fruit and vegetables'. In *Price and Market Policies in European Agriculture,* ed. K. J. Thompson and R. M. Warren. University of Newcastle upon Tyne.
Timmer, C. Peta, and Nesheim, Malden C. (1979) 'Nutrition, product quality and safety'. In *Consensus and Conflict in US Agriculture*, ed. Bruce L. Gardner and James W. Richardson. College Statim, Texas, Texas A & M University Press.
Timmer, C. Peta, Falcon, Walter P., and Pearson, Scott R. (1983) *Food Policy Analysis*. Johns Hopkins University Press, Baltimore.
Valdes, Alberto (ed.) (1981) *Food Security for Developing Countries*. Westview Press, Boulder, Colo.

The Food Consumer
Edited by C. Ritson, L. Gofton and J. McKenzie

CHAPTER 2

The Food People Eat[1]

DAVID BLANDFORD

The purpose of this chapter is to describe food consumption patterns in the industrialized countries, and to analyze how these patterns have been changing over the past two or three decades. It is based upon information for the countries of the Organization for Economic Cooperation and Development (OECD). These countries, listed in Table 2.1, are principally the richer countries of the world with an average per capita income of roughly 10000 US dollars in 1981.

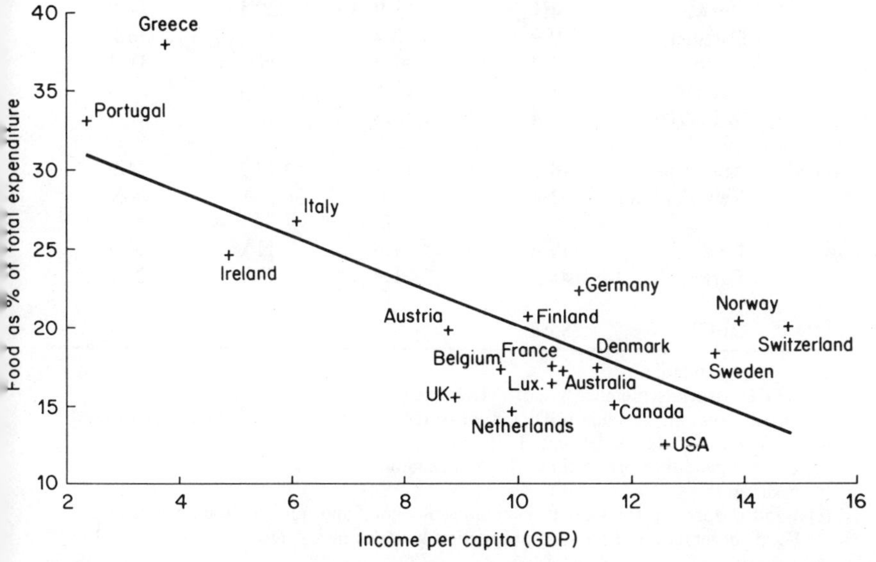

Figure 2.1 Relationship between income and food expenditure in the OECD in 1981.

They comprise most of Western Europe, North America and Australasia together with two Asian countries—Japan and Turkey.

The proportion of total consumer expenditure devoted to food varies among

Table 2.1 Selected characteristics of the OECD countries, 1981.

Region	Country	Population (Millions)	Income per capita GDP ($US 1000)	% of consumer expenditure on: Food	Food, beverages and tobacco
N. America	Canada	24.2	11.7	15.1	20.6
	USA	229.8	12.6	12.6	16.2
EEC—Europe	Belgium	9.9	9.7	17.4	23.3
	Denmark	5.1	11.4	17.5	25.1
	France	54.0	10.6	17.6	21.1
	Germany	61.7	11.1	22.4	24.3
	Greece	9.7	3.8	37.9	43.9
	Ireland	3.4	4.9	24.7	42.9
	Italy	57.2	6.1	26.8	31.0
	Luxembourg	0.4	10.6	16.5	20.4
	Netherlands	14.2	9.9	14.7	19.3
	UK	56.0	8.9	15.6	21.2
Other Europe	Austria	7.5	8.8	19.9	25.7
	Finland	4.8	10.2	20.8	27.3
	Iceland	0.2	12.8	NA	NA
	Norway	4.1	13.9	20.5	27.0
	Portugal	9.9	2.4	33.1	38.2
	Spain	37.7	4.9	NA	31.4
	Sweden	8.3	13.5	18.4	24.8
	Switzerland	6.4	14.8	20.1	27.6
Australasia	Australia	14.9	10.8	17.3	24.9
	New Zealand	3.1	8.0	NA	NA
Asia	Japan	117.6	9.6	NA	24.4
	Turkey	45.7	1.3	NA	NA
	OECD	785.8	9.7	16.3	21.1

Notes: NA = Not available.
(a) OECD associate member Yugoslavia is excluded from the table.
(b) Gross Domestic Product (GDP) is measured at 1981 prices and exchange rates.
(c) Expenditure data for Ireland, Italy and Spain are for 1980.
(d) Food expenditure proportions for Australia and Canada include non-alcoholic beverages.
(e) Food proportion for Germany includes alcoholic and non-alcoholic beverages. Food, beverages and tobacco proportion includes meals out.

Source: OECD *National Accounts Statistics*.

the OECD countries, ranging in 1981 from a low of 12.6 per cent in the United States to a high of 37.9 per cent in Greece.[2] Although data are not available on food expenditures in all cases, it is possible to demonstrate in Figure 2.1 that the proportion of income spent on food tends to decline fairly rapidly across the countries as income rises. In this figure, the food expenditure proportions from Table 2.1 are plotted against per capita income (Gross Domestic Product). The downward-sloping line on the graph is the average relationship between expenditure and income for the countries included. There are variations around the average due to differences in food preferences and price levels. For example, Norway and Sweden, which had levels of income similar to those of the USA in 1981, had substantially larger expenditure proportions. Most of the EEC countries are either close to or below the average relationship line. The principal exception, Germany, is due to the inclusion of alcoholic and non-alcoholic beverages in the German expenditure figure.

CHANGES IN TOTAL FOOD CONSUMPTION

Figure 2.1 suggests that, as consumers become wealthier, they tend to spend a declining proportion of their incomes on food. (This relationship is sometimes known as 'Engel's law' and it is discussed in the chapter by Tangermann.) In order to explore more fully how food consumption differs among countries and how patterns of consumption have been changing, it is necessary to go beyond expenditure figures and examine the quantity of food consumed. Cross-country comparisons based on monetary expenditures can be extremely misleading because of differences in price levels and changes in currency exchange rates.

Fortunately, the OECD makes available statistics on aggregate food consumption and its composition by product, expressed in calories per capita per day. The OECD figures relate to *apparent* net human consumption, allowing for waste in storage and processing. They do not necessarily represent actual consumption because there will be losses after processing, e.g. in the distribution system, in food preparation, and due to incomplete consumption by consumers themselves. Although the data may not be treated as precise estimates of consumption, they provide a useful basis for comparison across countries and an indication of changes over time.

The use of caloric figures makes it possible to aggregate different foods into a total, and to derive consumption shares by types of foodstuff. However, the use of these figures also imposes some limitations. A common set of conversion factors is used by the OECD for all countries and years to convert from product weight to calorie equivalents. The figures derived are an approximation to the *actual* coloric value of foods which may differ through time and across countries. The use of calories means that in aggregating commodities greater emphasis is given to foods which are high in calories per unit, e.g. fats and oils, cereals and starchy foods. Less emphasis is given to most meats, which are higher in proteins, and very little to most fruits and vegetables, which are low in calories per unit. The use of calories means that the assessment of changes in food consumption in this chapter will be dominated by those foods which are the principal sources of energy in the diet.

With these qualifying statements in mind, we may turn to Table 2.2, which contains figures for per capita daily food consumption for 21 OECD countries for 1960-1961 and 1979-1980.[3] It can be seen that average total food consumption for these countries in 1979-1980 was over 3200 calories per day and ranged from roughly 2600 calories in Japan to 3500 in the Netherlands.[4] The table also gives the average compound growth rate for total food consumption over the

Table 2.2 Per capita food consumption in the OECD countries.

Region	Country	Calories per day		Average annual growth rate %
		1960-1961	1979-1980	
N. America	Canada	2994	3084	0.2
	USA	3162	3367	0.4
EEC—Europe	Belgium/Lux.	3142	3282	0.2
	Denmark	3343	3455	0.2
	France	3124	3437	0.5
	Germany	3150	3423	0.5
	Ireland	3482	3520	0.1
	Italy	2726	3435	1.3
	Netherlands	3270	3480	0.3
	UK	3214	3228	0.0
Other Europe	Austria	3102	3333	0.4
	Finland	3109	3052	−0.1
	Norway	2974	3201	0.4
	Portugal	2640	3126	1.1
	Spain	2617	3073	0.9
	Sweden	2996	2997	0.0
	Switzerland	3327	3360	0.1
Australasia	Australia	NA	3139	NA
	New Zealand	3274	3107	−0.3
Asia	Japan	2302	2582	0.6
Mean		3050	3234	0.3
Average difference from mean (%)		7.2	5.4	

Notes: The most recent figures for New Zealand and Portugal are for 1976-1977, and 1978-1979 for the USA. Other OECD countries are omitted due to lack of data.
NA = not available.

Source: OECD *Food Consumption Statistics*.

period. In most countries, total food consumption increased, although there are some cases in which consumption remained stable or even declined slightly. The UK is one such case. Only three countries (Italy, Portugal and Spain) experienced a rate of growth in food consumption of roughly 1 per cent or more. In the remaining cases, total food consumption tended to expand at a more modest pace of 0.5 per cent or less since the beginning of the 1960s.

At the foot of the table, the average difference across countries from the mean level of consumption is given in percentage terms. Thus, in 1960-1961 average percentage variation around the mean consumption level of 3050 calories was just over 7 per cent. In 1979-1980 this number had fallen to less than 5.5 per cent. This decrease implies that the pattern of growth that has occurred since the early 1960s has tended to reduce the differences between the industrialized countries in terms of their total food intake. In this respect, at least, dietary patterns have tended to become more similar.

Conventional economic wisdom suggests that the principal determinants of changes in the consumption of a product are variations in consumer income and in the price of the product relative to its complements or substitutes. In the case of food, few other goods can be considered to be close substitutes. For this reason, it is likely that the principal economic determinant of long-term changes in per capita food consumption is the change in consumer income.

In Figure 2.2, food consumption data for three countries—Belgium, Germany

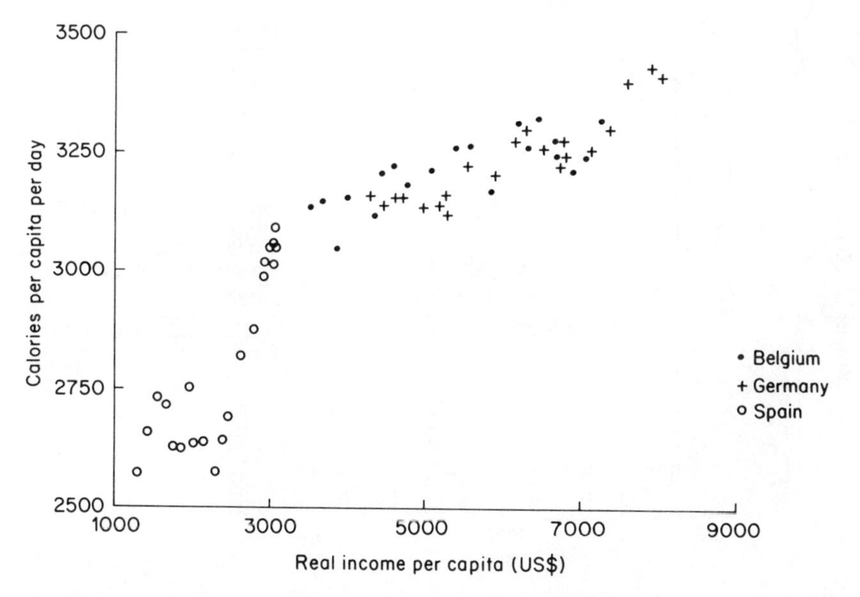

Figure 2.2 Relationship between food consumption and income in selected OECD countries for 1960-1980.

and Spain—for 1960-1980 are plotted against their real per capita income (per capita GDP in US dollars at 1975 prices and exchange rates). This figure suggests that a positive relationship exists between changes in the level of income through time and total food consumption in these countries. As income rises, total food consumed tends to increase also. An analysis of similar graphs for other countries suggested that just over half of those listed in Table 2.2 displayed such a relationship. However, in the case of Australia, Canada, Denmark, Finland, Ireland, New Zealand, Switzerland and the UK, no consistent relationship appeared to exist between changes in income and food consumption over the decades of the 1960s and 1970s. In these countries, food consumption has apparently reached a plateau and no longer displays any consistent trend.

For the remaining countries, the relationship between income and food consumption can be estimated statistically. In making such estimates, it was assumed that food consumption will not increase indefinitely as incomes rise but, as in the case of the countries listed above, will eventually reach a plateau. The methods used to estimate the relationship are described in more detail elsewhere (Blandford, 1984). However, their implications can be determined from the summary of results given in Figure 2.3.

This chart shows the number of OECD countries whose likely consumption response to changes in income may be classified as high, medium or low in

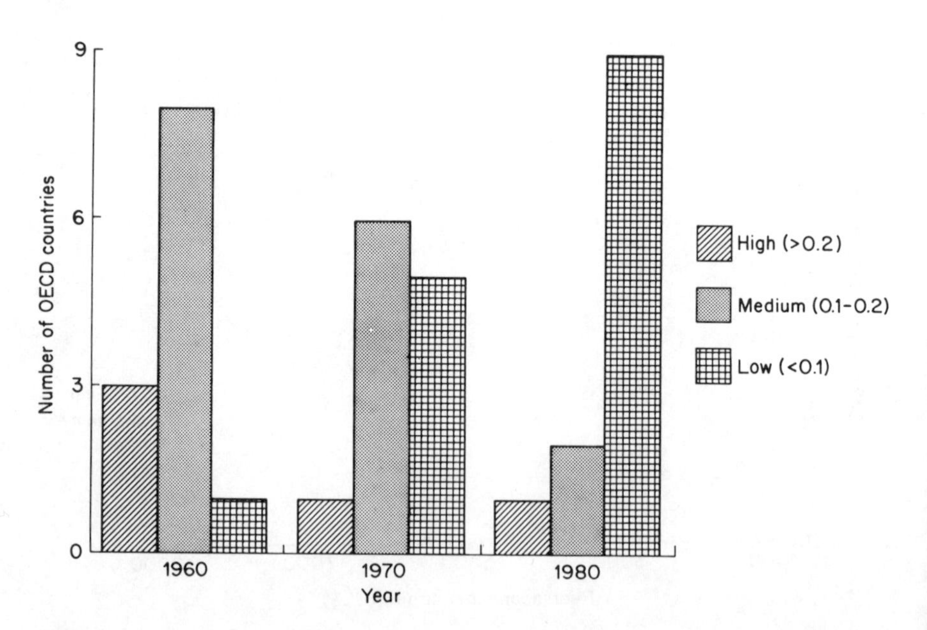

Figure 2.3 Responsiveness of total food consumption to changes in income (Blandford, 1974).

selected years. For the high response countries, a 1 per cent increase in income is associated with a 0.2 per cent increase in total food calories consumed per capita. In the medium category of countries, this response would be between 0.1 and 0.2 per cent, and in the low category, response would be positive but less than 0.1 per cent. Only twelve countries are included in the chart since eight had no apparent relationship between income and total food consumption for the period analyzed. As the bars in Figure 2.3 show, the responsiveness of total food demand to changes in income in the industrialized countries has been progressively diminishing through time. In 1960 there were eleven countries that could be classified as moderately or highly responsive. By 1980 only three countries fell into these categories. It should come as no surprise, given the information already presented in Table 2.2, that the countries which still demonstrate some responsiveness to the growth in income are Italy, Spain and Portugal. However, even these countries have shown a sustained reduction in the magnitude of this response. In Spain, for example, the estimates derived suggest that an increase in real per capita income of 1 per cent in 1960 would have generated an increase in per capita food consumption of just over 0.5 per cent. By contrast, in 1980 the same percentage increase in income would have generated an increase in food consumption of 0.2 per cent.

The conclusion to be drawn from these results is that per capita food consumption in most of the richer industrialized countries has either reached or is rapidly reaching a plateau. Although the level of this plateau may differ from country to country, and for some countries the potential still exists for further growth as a result of increases in income, the likelihood of further significant expansion in per capita food consumption in most industrialized countries is small. In fact, the estimates derived suggest that in the future total caloric consumption of food could actually decline in some countries. Since population is also growing relatively slowly in most OECD countries, it appears likely that aggregate food consumption in the industrialized world will grow slowly in the future, even if rates of growth in per capita income were to recover from the depressed levels of the past few years.

CHANGES IN THE CONSUMPTION OF ANIMAL VERSUS VEGETABLE PRODUCTS

One of the major characteristics of the diet is its composition in terms of animal products, such as meat, milk and eggs, and vegetable products, such as grains and flour, roots and tubers, and fruits and vegetables. Table 2.3 provides information on the level of per capita consumption of animal products in calories in 1960-1961 and 1979-1980 for the countries discussed above.

The table indicates that per capita consumption of animal products in 1979-1980 averaged over 1300 calories per day, ranging from roughly 560 calories in Japan to 1700 calories in Denmark. In most countries, the total consumption

Table 2.3 Per capita consumption of animal products in the OECD countries.

Region	Country	Calories per day 1960-1961	Calories per day 1979-1980	Average annual growth rate %	Animal calories as % of total 1960-1961	Animal calories as % of total 1979-1980
N. America	Canada	1366	1458	0.4	45.6	47.3
	USA	1331	1262	−0.3	42.1	37.5
EEC—Europe	Belgium/Lux.	1135	1456	1.4	36.1	44.4
	Denmark	1522	1718	0.7	45.5	49.7
	France	1128	1546	1.8	36.1	45.0
	Germany	1255	1568	1.2	39.8	45.8
	Ireland	1441	1492	0.2	41.4	42.4
	Italy	530	946	3.3	19.4	27.5
	Netherlands	1076	1432	1.6	32.9	41.1
	UK	1150	1306	0.7	35.8	40.5
Other Europe	Austria	1130	1492	1.6	36.4	44.8
	Finland	1197	1464	1.1	38.5	48.0
	Norway	972	1210	1.2	32.7	37.8
	Portugal	498	759	2.8	18.9	24.3
	Spain	410	917	4.6	15.7	29.8
	Sweden	1263	1380	0.5	42.2	46.0
	Switzerland	1159	1448	1.2	34.8	43.1
Australasia	Australia	NA	1164	NA	NA	37.1
	New Zealand	1643	1558	−0.4	50.2	50.1
Asia	Japan	222	559	5.3	9.6	21.6
Mean		1075	1307	1.1	35.3	40.4
Average difference from mean (%)		26.9	17.8		22.3	16.1

Notes: The most recent figures for New Zealand and Portugal are for 1976-1977, and 1978-1979 for the USA. Other OECD countries are omitted due to lack of data.
NA = not available.

Source: OECD *Food Consumption Statistics.*

of animal products has increased since the early 1960s. In twelve of the countries listed, the average rate of increase exceeded 1 per cent per year. In several cases, most notably Italy, Portugal, Spain and Japan, the increase in consumption was particularly rapid. On the other hand, in two cases, New Zealand and the United States, the number of calories derived from animal products tended to decline. In both cases this is largely due to a reduction in the consumption of butterfat.

The rapid growth in the consumption of animal products, coupled with the sluggish growth of total food consumption, has meant that in most OECD countries the share of total food calories derived from animal products has increased substantially. Animal products have, therefore, tended to replace vegetable products in the diet. Only in the United States was the share of calories derived from animal products significantly lower at the end of the 1970s than at the beginning of the 1960s. For the OECD countries as a whole, the proportion of animal calories rose from roughly 35 per cent to over 40 per cent during the period.

The figure for the average difference from the mean of 16 per cent at the foot of the table demonstrates that considerable variation still exists in the significance of animal products in the diet. However, as in the case of total food consumption, this difference has tended to diminish. Hence, we may conclude that there is an increasing similarity among the industrialized countries in the volume of animal products consumed per capita, and in the relative significance of these products in the diet.

As in the case of total food consumption, it is to be anticipated that change in the level of income is a major determinant of long-term change in the proportion of food derived from animal sources. In addition, it is likely that changes in the price of animal products relative to vegetable products have had an impact. The decline in the relative price of many animal products which has been associated with the growth of industrialized livestock production in Europe and North America may have reinforced the trend towards increased consumption of animal products as the result of income growth.

For a few countries (Australia, Canada and New Zealand), an examination of graphs of the animal product share and per capita income over the period 1960 to 1980 revealed no consistent relationship between the two. For fourteen countries, a positive relationship with income was apparent, while for the United States a negative relationship appeared to exist. For the two remaining countries, Ireland and the UK, a noticeable change in the relationship between the animal product share and income has occurred. In the earlier part of the period, a positive relationship existed, but since the early 1970s negative association is apparent. This could be due to a change in food preferences or, more probably, to the effect of the increase in the price of animal products in these two countries following their accession to the European Community.

The typical relationship which exists between income and animal product consumption in the OECD countries is illustrated in Figure 2.4. This is based upon the same countries included in Figure 2.2—Belgium, Germany and Spain—and suggests that there is a general tendency for the significance of animal products in the diet to increase as income rises, although at some point the proportion of animal calories consumed tends to reach a plateau.

As in the case of total food consumption, the responsiveness of animal product consumption to income can be estimated statistically. Figure 2.5 is based upon the results obtained (Blandford, 1984). Countries are divided into high, medium,

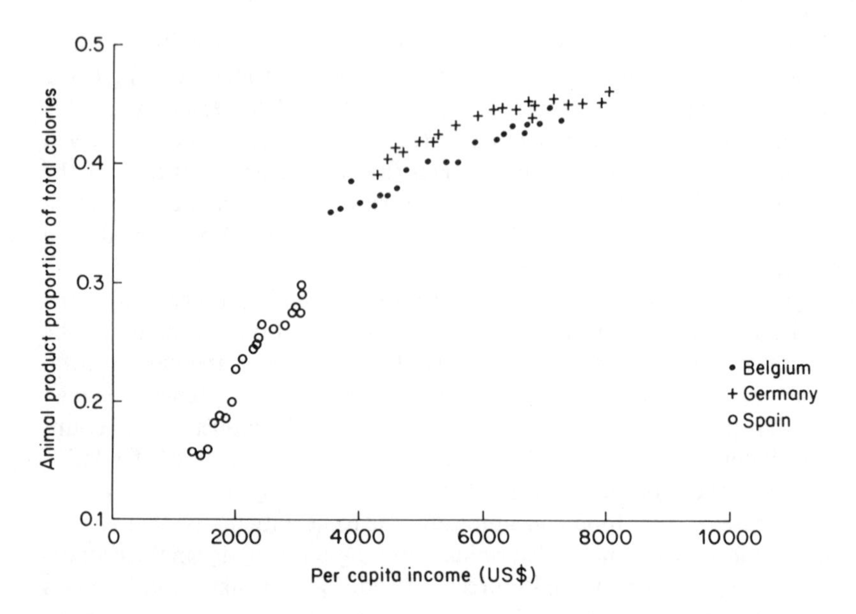

Figure 2.4 Relationship between animal product consumption and income in selected OECD countries for 1960-1980.

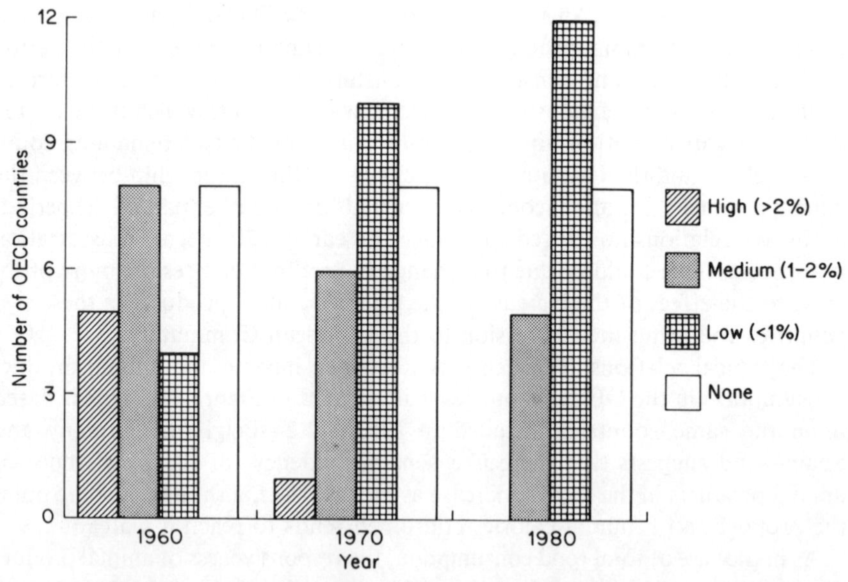

Figure 2.5 Response of animal product consumption to changes in income (Blandford, 1984).

low and no response categories. High response implies, for example, that a 10 per cent increase in per capita income would be associated with an increase of at least 2 percentage points in the share of animal calories in the diet. From Figure 2.5 it can be determined that the responsiveness of animal product consumption to changes in income has tended to fall through time in the OECD area. In 1960 more than half of the countries analyzed could be classified as having high or medium response to changes in income. Increases in the animal product share of 1 percentage point or more would have occurred in thirteen countries, with Spain, the UK, Italy, Japan and the Netherlands having increases of over 2 percentage points. In contrast, by 1980 no countries remained in the high response category and a 10 per cent rise in income would have led to an increase of 1 percentage point or more in the consumption of animal products only in Italy, the Netherlands, Spain and Sweden. The majority of the OECD countries in this year had the potential for either little or no response to changes in income in the proportion of animal products consumed. In general, the results indicate that further expansion in the share of animal products in the total food consumed is possible with income growth in the OECD countries, although the rate of such potential expansion is falling.

CHANGES AND SIMILARITIES IN DIETARY PATTERNS

An examination of total food consumption and its animal product component enables us to summarize the essential characteristics of dietary structure in the industrial countries and to explore the way in which this may change in the future. For example, Table 2.4 presents a classification based on the figures for 1979-1980 for the two major characteristics of total caloric intake and its animal product proportion. In this table, high, medium and low categories are identified for both characteristics. Total food consumption is viewed to be high if it exceeds 3200 calories, medium if it ranges between 2800 and 3200 calories, and low if it is less than 2800 calories per day. Animal product consumption is classified as high if it exceeds 35 per cent of total food intake, medium if it ranges between 25 and 35 per cent, and low if it is less than 25 per cent.

On the basis of the criteria used, the bulk of the OECD countries have high animal product consumption and either high or medium levels of total food intake. Only Spain, Portugal and Japan fall in the medium or low categories. In the case of the European countries, this may be related to cultural factors and income levels. Physiological factors are probably important in the case of Japan.

Table 2.4 reinforces the point that substantial similarity in dietary structure exists in the industrialized world. This is further confirmed by Figure 2.6 which presents a more detailed picture of dietary patterns in OECD countries. This chart is based upon the statistical analysis of similarities in caloric shares per country for eight major food groups: cereals and starchy foods; meat and eggs;

Table 2.4 Classification of OECD countries by total food consumption per capita and its animal product component, 1979-1980.

Animal product consumption % of total calories	Total food consumption (calories) High (>3200)	Medium (2800-3200)	Low (<2800)
High (>35)	Austria, Belgium/Lux. Denmark, France, Germany, Ireland, Netherlands, Norway, Switzerland, UK, USA	Australia, Canada, Finland, New Zealand, Sweden	
Medium (25-35)	Italy	Spain	
Low (<25)		Portugal	Japan

Source: OECD, *Food Consumption Statistics.*

oils and fats (including butter); sugar; milk and milk products (excluding butter); fruits and vegetables; pulses and nuts; fish. Country aggregates were derived on the basis of the overall similarity of consumption shares across this set of products. (See Blandford, 1984, for details.) In the process, the greatest weight

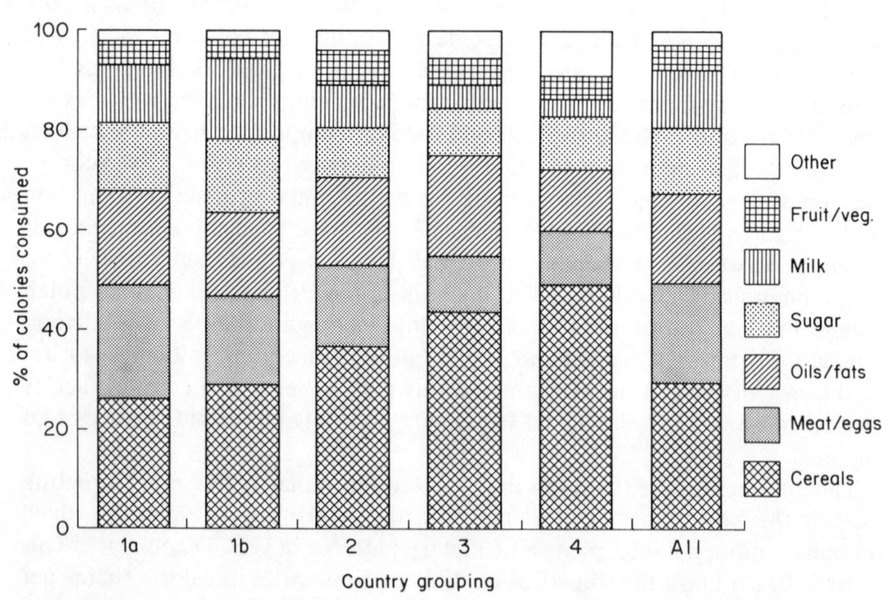

Figure 2.6 Dietary differences in the OECD area in 1977-1979 (Blandford, 1984).

was placed upon the degree of similarity in major caloric components, typically the first five food groups defined above. Relatively little weight was placed upon differences in the proportion of total calories obtained from the less significant caloric sources such as fruits and vegetables, pulses and nuts, and fish.

The groupings presented in Figure 2.6 reinforce the impression that considerable similarity in the structure of the diet exists across the OECD countries. Group 1a, which has the lowest proportion of total calories obtained from cereals, contains twelve countries, including the OECD members in North America and Australasia plus six members of the European Community (Belgium and Luxembourg, Denmark, France, Germany and the Netherlands). These countries are all characterized by a low cereals component and by high meat and eggs and oils and fats components in the diet. Group 1b, which contains the Scandinavian countries (Finland, Norway and Sweden) plus two EC members (Ireland and the UK), has a dietary structure broadly similar to that of group 1a with the exception that it obtains a substantially larger share of its calories from milk products and a substantially smaller share from meat and eggs.[5]

Groups 1a and 1b combined include 17 of the 21 OECD countries analyzed. The remaining four countries are divided into three groups. Group 2 contains Italy and Spain. These two countries obtain a significantly higher proportion of their calories from cereals than group 1 and significantly lower proportions from sugar and milk. The proportion of calories obtained from meat and eggs and from oils and fats by Italy and Spain is, however, similar to that of group 1b.

Group 3 contains Portugal alone. This country obtains a substantially larger proportion of its calories from cereals than the other Mediterranean countries—Italy and Spain—and substantially less of its calories from meat and eggs and from milk. Group 4 contains Japan alone, and differs from Portugal in the substantially lower proportion of total calories obtained from oils and fats and the higher proportion obtained from cereals.

For the OECD countries as a whole, only 30 per cent of total food calories are obtained from cereals and starchy foods. The degree to which this proportion, and the proportion of calories derived from other food groups, has changed over the past two or three decades is illustrated by Figure 2.7. In the mid 1950s, 40 per cent of total calories were obtained from cereals in the OECD area. A 10 per cent reduction in this proportion by the late 1970s was primarily taken up by a 5 per cent increase in the proportion of calories derived from meat and eggs and a 3 per cent increase in oils and fats. Sugar and milk products have increased their share of the diet by roughly 0.5 per cent each and fruits and vegetables by 1 per cent. As cereals and starchy foods have been displaced by more varied and more expensive sources of calories, practically all other food groups have increased in importance in the consumer's diet in the industrialized countries.

Will this change in dietary patterns continue? Although it is difficult to answer this question using the level of detail contained in Figure 2.7, it is possible to use

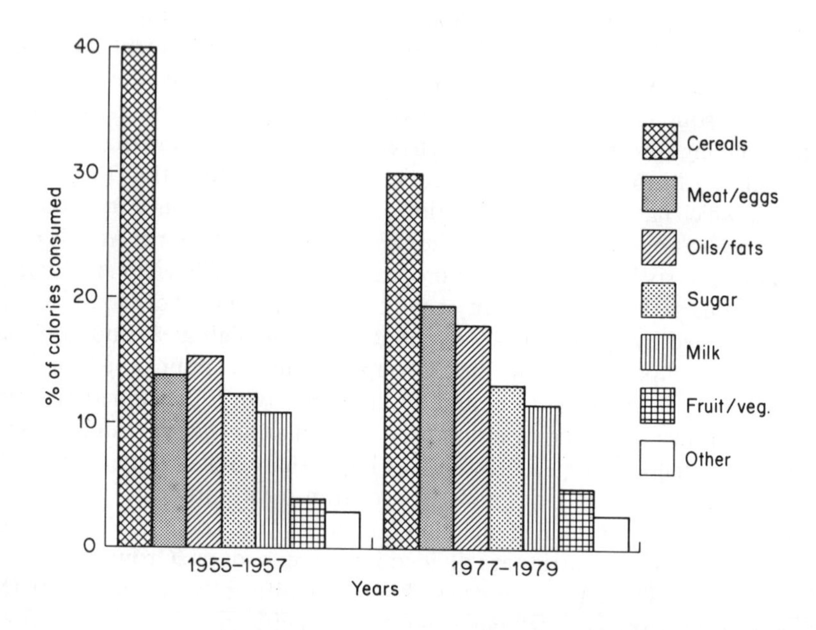

Figure 2.7 Changes in dietary structure in the OECD area (Blandford, 1984).

the statistical results referred to earlier in this chapter to examine likely future changes in total food consumption and in animal versus vegetable product consumption in the industrialized countries. Table 2.5 summarizes the implications of the results of the statistical analysis for food consumption in each country with further growth in per capita income. It indicates that there are six possible types of relationship between changes in total food consumption and in the consumption of animal and vegetable products. The results summarized in Figures 2.3 and 2.5 suggest that income growth is likely to generate little change in per capita consumption of food as a whole, and in the per capita consumption of animal or vegetable products in three countries—Australia, Canada and New Zealand. On the other hand, although little change in aggregate per capita food consumption is likely in a further five countries, in three of these (Denmark, Finland and Switzerland) further substitution of animal products for vegetable products would probably occur, and in Ireland and the UK it is possible that vegetable products would substitute for animal products.

Growth in per capita income would probably lead to an increase in aggregate food consumption per capita in thirteen countries. In seven of these, the increase would be accompanied by the substitution of animal calories for those derived from vegetable products (Austria, Belgium and Luxembourg, France, Italy, Japan and the Netherlands). In the United States, the reverse would occur, i.e. the substitution of vegetable for animal calories. In the remaining five countries

Table 2.5 Probable changes in per capita food consumption with the growth in per capita income.

	No increase in either animal or vegetable products	Substitution of animal for vegetable products	Substitution of vegetable for animal products	Increase in both animal and vegetable products
No change in per capita total food consumption	Australia Canada New Zealand	Denmark Finland Switzerland	Ireland UK	NA
Increase in per capita total food consumption	NA	Austria Belgium/Lux. France Italy Japan Netherlands	USA	Germany Norway Portugal Spain Sweden

NA = Not applicable.
Source: Blandford (1984).

(Germany, Norway, Portugal, Spain and Sweden), both animal and vegetable product consumption would probably increase as the result of income growth.

CONCLUDING COMMENTS

In recent decades, food consumption patterns in the industrialized countries have changed substantially. However, as incomes have increased, per capita food consumption has tended to stabilize, and although growth is still possible in some countries, this is likely to be small. The per capita consumption of food calories could actually decline in some countries in the future as food preferences continue to change towards lower calorie foodstuffs. Since population is also growing relatively slowly in most richer countries, it appears likely that aggregate food consumption will experience only modest growth in the future even if incomes were to expand rapidly.

In virtually all countries, the growth of income in the past 30 years has been associated with a decline in the importance of vegetable products, such as cereals and starchy foods, in the diet and an increase in the importance of animal products such as meat and eggs. However, in most richer countries the share of animal products in total food consumption is tending to stabilize, as its responsiveness to income growth declines. It is, therefore, unlikely that the historical trend of rapid and regular increases in the share of animal products in total food consumption will continue indefinitely in these countries. Increasing concern about the health and nutritional implications of high levels of consumption of animal fats is likely to accentuate this trend.

Overall, dietary patterns in the industrialized world have tended to become increasingly similar despite substantial variation in the proportion of consumer expenditure devoted to food. Although dietary differences are likely to persist due to cultural or physiological factors, most countries appear to be adopting a diet which is strikingly similar in terms of the importance of the major food groups. The 'typical' diet is characterized by a low share of cereals and starchy foods compared to consumption in low-income countries, and high shares of meat and eggs, oils and fats, and milk and milk products. Although it unlikely that this pattern will remain unchanged as we approach the twenty-first century, the probability of major structural changes would seem to be small.

NOTES

1. This chapter is based on the author's article, published in the *European Review of Agricultural Economics.* The editors are grateful to the managing editor of the *Review* for his permission to use the article in this way.
2. These expenditure proportions do not include food consumed away from home, e.g. purchased in restaurants.
3. Throughout the remainder of the chapter, where groupings of countries are indicated Belgium and Luxembourg are treated as a single entity. The remaining three OECD countries (Greece, Ireland and Turkey) are omitted because of the lack of suitable data.
4. Since the averages given in this and subsequent tables are not weighted by population, they are not actual per capita figures for the sum of the countries listed.
5. As indicated elsewhere (Blandford, 1984), the substantial physical similarity in dietary structure for the bulk of the OECD countries represented in these two groups exists despite considerable variability in the proportion of consumer expenditure devoted to food. Such variability is created both by differences across countries in the level of income and differences in relative prices.

REFERENCES

Blandford, D., (1984) 'Changes in food consumption patterns in the OECD Area', *European Review of Agricultural Economics,* **11**, 43-64.

OECD. *Food Consumption Statistics,* various issues, Paris.

OECD. *National Accounts Statistics,* various issues, Paris.

The Food Consumer
Edited by C. Ritson, L. Gofton and J. McKenzie

CHAPTER 3

Food and Nutrition

Principles of Nutrition and some current controversies in Western countries

ARNOLD BENDER

THE MEANING OF NUTRITION

The systematic study of nutrition dates only from the beginning of this century—the word 'vitamin' was coined as recently as 1912. The study and its application rest upon a wide range of other academic disciplines. My characterization of this relationship is shown diagrammatically in Figure 3.1.

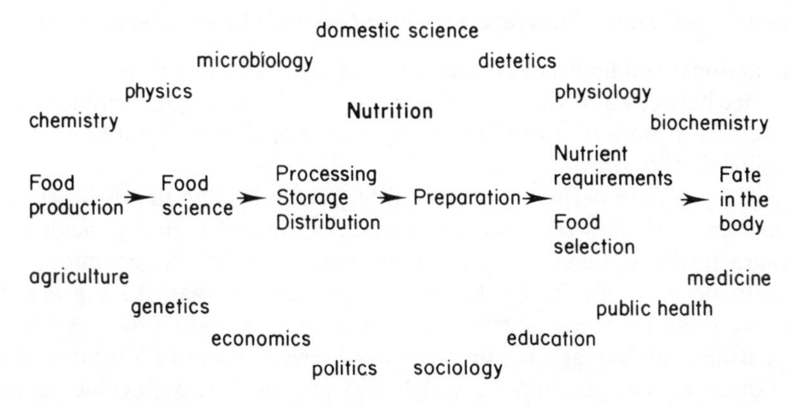

Figure 3.1 Relationship between nutrition and other academic disciplines.

Several of the surrounding areas of study are covered by other chapters of this book. The present chapter is confined to the effect of food upon the human body. It deals first with basic problems in considering the current state of nutrition, which are mainly problems of measurement. Then the chapter explores the many aspects of the changing foodstuffs that technological and social progress have brought into the diet of consumers in industrialized countries. Finally, it looks towards the food and nutrition of the future.

CURRENT STATE OF NUTRITION—BASIC PROBLEMS

The most important basic question the nutritionist has to answer falls into three parts.

(1) What constitutes a good diet;
(2) how do we determine whether individuals are consuming such a diet; and,
(3) how can one rectify poor dietary habits?

Over the past century the essential dietary factors have been determined. It took 50 years to discover that 14 vitamins, and somewhat longer to find that 21 mineral salts, are essential in the diet (and it is possible that this list is not yet complete).

Recent experience of feeding surgical patients intravenously with solutions of the required nutrients, a procedure which assisted the discovery of some of the essential trace minerals, suggests that we do know what nutrients are required, at least in the short term. There may, however, be others, deprivation of which does not show any effects during short periods of intravenous feeding. Even then we have not answered the question of how much of each is necessary to maintain health.

Recommended Daily Amounts or Allowances (RDA) or Intakes (RDI)

Many national and international authorities have drawn up tables of RDAs (the difference between allowances, amounts and intakes is of philosophical rather than scientific import) intended to apply to population groups but not to individuals.

These figures are derived by one of three methods. The first is observation of the dietary intakes of specific nutrients, compared with the incidence of the characteristic deficiency disease. Thus, when beri-beri was common, it was found to occur when the intake of thiamin was less than 0.5 mg per day. Since the body becomes saturated at intakes of 1 mg (the excess is excreted in the urine) the average requirement lies between these two figures. There are, however, few nutrients to which this approach is applicable, because many deficiency diseases are not so specific or may take a long time to develop.

The second method is animal experimentation. Large numbers can be used, being fed with varying dose levels to ascertain the therapeutic dose with some precision. The drawback lies in the problem of extrapolating from these results to human beings. This can be done on a sound basis only when the mode of action of the nutrient in the body is fully understood.

In the case of beri-beri, thiamin functions as a co-enzyme in the metabolism of carbohydrates and proteins, and the amounts required by both animals and man are best expressed in milligrams per 1000 non-fat calories (0.5 megajoules). The disease occurs when dietary intakes of thiamin fall below 0.24 mg per 0.5 MJ.

Other nutrients are usually scaled up in terms of the relative amounts of actively metabolizing body tissue, i.e. body weight excluding fat reserves and skeleton.

The third approach, which for obvious reasons is very rarely used and then only on few subjects, is direct human experimentation. This method was used to determine the requirements of vitamin C and vitamin A in Sheffield over the years 1942-1944. Twenty-six volunteers lived on diets complete in all respects except for the vitamin under investigation, and various blood parameters and enzyme functions were measured at intervals. When clinical signs of deficiency occurred, the curative dose could be determined.

Clinical signs of vitamin C deficiency appeared after 5 months, but only a few of the subjects showed signs of vitamin A deficiency, even after 2 years. This is a measure of the reserve stocks held in the liver of a well-fed individual.

The figures derived by the three methods agree reasonably well for some of the nutrients, but even for these, two major problems remain, both vital to the study of nutrition. One is that the figures relate to the amounts required to cure or prevent a deficiency; they do not provide information about optimum intakes. The second is that individuals vary in their needs.

The optimum intake is often a matter of opinion and different criteria may be used. Vitamin C illustrates some of the problems. The Sheffield experiments showed that 5-10 mg cured scurvy but that wound healing was delayed; 20 mg was necessary for proper wound healing. To allow for individual variations, a generous average brought the figure to 30 mg per day, which is the figure adopted by Britain (and the FAO).

A different criterion was used in the United States. Most animals are able to synthesize their own vitamin C, so it is reasonable to conclude that their blood levels are 'correct'. To achieve similar levels in human beings requires an intake of 75 mg, the figure adopted by the US authorities in 1945. The poverty of the evidence and the latitude of opinion are illustrated by the fact that the US figures have ranged over the years from 75 to 60 to 45 and, currently, back to 60 mg per day.

The problem of individual variation is taken care of statistically. The average requirement plus two standard deviations should cover the needs of those at the high end of the scale and this is the figure used in tables of RDAs which are used to monitor average nutrient intakes in nutrition surveys and to plan food supplies. It can be assumed that if an individual is consuming as much as the RDA then his needs are well catered for.

Nutritional Status

It is difficult to assess the nutritional status of an individual. Even poorly fed people cannot be diagnosed simply from clinical examination, since their signs are not always characteristic nor very specific.

Further information is sought from biochemical measurements on blood and tissues while dietary analysis may assist by indicating a low intake of the suspected nutrient. The final proof is obtained if the subject responds to increased intake of the specific nutrient.

This applies to cases of severe malnutrition. It is much more difficult, if at all possible, to diagnose marginal malnutrition.

Dietary Intake

Even if we were able to determine what a particular individual required, we still have the problem of measuring the amount of each nutrient the subject is consuming. The most reliable method is to weigh all foods eaten and to analyse duplicate samples for their nutrient content. This is a very lengthy and laborious procedure and therefore rarely carried out. It is more usual to weigh the foods and to calculate the nutrients from tables of food composition. Errors obviously arise, since the foods consumed differ from those originally analysed when the tables where constructed. For large numbers of subjects, even less accurate methods must be used, such as estimating portion sizes or regular questioning by a dietitian.

It is clear that the nutritionist is still seeking the answers to the most fundamental questions of the discipline.

A problem, which at the same time is a safeguard, is that human beings adapt to variations in their nutrient supplies. Our Western, i.e. scientifically determined, figures show that a heavy manual worker needs about 3000 kcal (12 MJ) per day, yet peasant-farmers engaged in heavy work manage on less than 2000 kcal. In developing countries, mothers have been found capable of breast feeding their infants, yet while the infants are growing and the mothers maintain their body weight, their total food intake is only two-thirds of that thought to be necessary.

What is Good Nutrition?

Improved nutrition, together with many other environmental factors, has led to the earlier attainment of maturity as measured by growth rates and age of menarche. During the same period there has been an increase in the development of degenerative diseases, which raises the question as to whether one leads to the other.

It is axiomatic that poor diets result in slow growth; slow growth is often due to malnutrition. Consequently, it would seem that fastest growth is 'best'.

This is not necessarily true. On the contrary, experiments dating from 1937 on a variety of animals show that underfed animals live longer than their well-fed litter mates. More recently it has been shown that, when treated with carcinogens, underfed animals develop fewer tumours than well-fed animals.

So, while we are still unable to answer the question as to what is meant by 'a good diet', there is evidence to suggest that current diets play a part in some of the diseases of Western society and that it is possible to improve such diets.

Dietary Goals

A major change in the disease patterns of Western countries has taken place over the past few generations. The major infectious diseases—tuberculosis, enteric fever, infectious diseases of childhood—have been so reduced that they no longer pose a major threat to life, while infant mortality has been dramatically reduced. The reasons are connected with a number of social and economic changes as well as changes in health services and environmental improvements of many kinds. Better nutrition has played some, undetermined, part in this, but in one specific area, namely the disappearance of nutritional deficiency diseases, the picture is clear.

The most common causes of death today in Western countries are heart disease, cancer and strokes. In Britain they account for 66 per cent of all deaths and for 72 per cent of deaths occurring after the age of 65. These, together with diabetes, diverticular disease and some others, are stigmatized as 'diseases of affluence' because they are common in richer, industrialized countries and comparatively rare in poor, undeveloped countries. Since their increase has partly coincided with changes in food habits, diet has been called into question. As a result, a vast amount of research has been carried out, mostly, unfortunately, without clear results.

The problems of ascertaining the causes of these diseases are exemplified by the finding that in coronary heart disease there are three groups of risk factors. The first consists of the unchangeable inborn factors, namely family history, maleness (it is more common among men than women until after the menopause), endomorphic body build, and certain behavioural factors and personality traits such as aggressiveness and drive.

The second group consists of a number of abnormalities that merge into disease states and include high blood lipid levels (hyperlipidaemia), hypertension, diabetes, obesity, high blood levels of uric acid and abnormal electrocardiogram patterns.

Only the third group of risk factors, namely the cultural and environmental factors, are open to change; these include smoking, lack of exercise, emotional stress and a number of nutrients. The latter are the ones that are the subject of the present discussion and they apparently play only a small part among the large number of factors. Hence the difficulty of 'proving' that specific dietary changes can be effective.

Only one dietary factor has met with general agreement, namely that the current Western diet contains 'too much' fat. Just how much is optimal is not known, nor is it likely that any fixed value could ever be regarded as ideal, since there must be a range of acceptable intake levels. When fat is less than about 10 per cent of the total intake, the energy density is so low that sheer bulk prevents an adequate intake of energy. More important, perhaps, is that, at this level of fat, the fat-soluble vitamins are poorly absorbed.

The level of fat in Western diets that is regarded as being 'too high' is when it is over 40 per cent of energy intake: the 'best' level is regarded as somewhere between 10 and 40 per cent. There is widespread agreement that the level of fat should be reduced to about 30-35 per cent of the diet; to reduce it below this level would call for such major changes in food habits that they would probably be unacceptable to consumers.

Apart from the association between a number of diseases and high fat intake, it can be shown experimentally that certain saturated fatty acids increase the levels of blood cholesterol, a known risk factor in heart disease. Polyunsaturated fatty acids from vegetable and marine sources can reduce blood cholesterol levels, but since an increase in their consumption militates against a decrease in total fats, there is no general agreement that such fats should be increased.

Among the dietary changes that have accompanied changing disease patterns is increased consumption of sugar—which has reached 20 per cent of the total energy intake in Great Britain. Sugar is an essential part of a large number of processed foods, as well as being a flavour enhancer in smaller amounts in many others, so that the visible sugar intake, e.g. the sugar intentionally added to drinks, is only half the total sugar intake.

High salt intakes are associated with hypertension in genetically susceptible individuals (possibly 15 per cent of the population) and with strokes.

Consequently, nutritionists have drawn up a list of dietary goals for which there is general, although not universal, agreement (NACNE, 1983; DHSS, 1978):

(1) Overweight persons should reduce their weight to 'normal' levels.
(2) Fat intake should be reduced from the current level of 40 per cent to 30-35 per cent of total energy intake. Some authorities advocate increasing the intake of polyunsaturated fatty acids.
(3) Reduce sugar intake.
(4) To compensate for the reduction in fatty and sugary foods, increase the intake of starchy foods such as cereals and potatoes.
(5) To increase the intake of dietary fibre, increase the consumption of fruit and vegetables, and replace white, milled cereals, particularly white bread, with whole grains and wholemeal bread.

These changes would assist weight control since these foods are less energy-dense than the sugary and fatty foods that they replace.

(6) The role of dietary cholesterol in raising blood cholesterol levels has been the subject of much investigation. While it has little effect on blood levels, some authorities recommend reducing the dietary intake to about 300 mg per day. The British authorities do not consider that the current daily intake of 350-450 mg is excessive and do not make any specific recommendations.
(7) Finally, salt intake should be reduced.

Such nutritional advice is recommended for the entire population. If it is true that high average national intakes of fats, sugar and salt are harmful, then the average *national* health should improve by reducing their consumption. What has not been proven is that such a change will benefit any individual directly. On the contrary, there are many people whose diet already follows these precepts, and indeed whose lifestyle avoids many of the risk factors listed above, yet who succumb to heart disease while still young. Similarly there are many individuals who flout all these tenets and live a long and healthy life. That is the dilemma. Can we offer advice as vague as that indicated, when the public assumes that all the dangers should be attributed to the next person? Do we wait until proof of the causes of these diseases is available—and many research workers argue that direct proof will never be available—or do we use the knowledge that we have and offer this advice?

In practice, the latter course is being adopted in many countries. There has been a fall in the incidence of heart disease in the United States and in some other countries, and credit is claimed by all those with vested interests—changes in exercise, in smoking and in fat, sugar and salt intake (Anon, 1980). On the other hand we know that over the years certain diseases have lost their virulence for no known reason and any national fashions such as marathon running and dietary changes may be simply accompanying changing disease patterns rather than causes of change. The next decade should answer some of these questions. If the dietary recommendations do indeed achieve their objective, then they will have proved themselves. Persuading the public to carry out the changes presents another problem.

CHANGING FOODS

Factory-produced foods have been replacing home-prepared foods steadily over the last couple of generations and give rise to the question as to whether such foods are 'as good as those mother used to make'.

Major developments in food production and marketing took place in all industrialized countries immediately after the Second World War. There were several reasons, which complemented and reinforced one another, for these developments. Many housewives went out to work, which both generated the demand for convenience foods, and provided the money with which to buy them. The chemical industry had demonstrated through the war years the valuable role of the chemical laboratory, and this was adopted by the food industry for laboratory control of raw materials, processing techniques and quality control of finished products. There was a rapid development of food science—Europe, following America, established university courses and concomitant technological courses in food science. This was accompanied by major research in food science and the development, in food engineering, of new techniques and new types of machinery, which in turn led to large-scale production and the standardization of products.

The new products and marketing techniques in turn accelerated the demand for them. This was accompanied by the introduction of new devices for use in the home, ranging from simple electrical gadgets in the earlier days, to refrigeration, deep freezers and microwave ovens. Large-scale production and international marketing led to multinational companies selling the same products in several countries (although there are still important differences in local tastes).

Then came the almost inevitable backlash, epitomized in Great Britain by the Campaign for Real Ale, the Campaign for Real Bread, and a variety of so-called health movements towards organic, natural, untreated and unprocessed foods. The size of these movements and their relative sincerity and scientific basis vary enormously from one country to another. There is confusion in the minds of the public over such terms as 'convenience foods', 'junk foods', 'natural', 'refined', etc. As more factual information reaches the public (and it must be borne in mind that the science of nutrition is relatively new, and attempts to inform the public are of even more recent origin) the problems should sort themselves out.

Nutritional Effects of Food Processing

The effects of processing, meaning the differences between home-prepared foods and factory ready-made convenience foods, need to be viewed in perspective (Bender, 1978).

Some nutritional losses are intentional

For example, when wheat is milled into white flour, or brown into white rice, the intention is to discard parts of the cereal not wanted by the consumer—with consequent simultaneous discarding of a proportion of the nutrients. Such processing losses may simply be a response to consumer demand.

Some losses are inevitable

When foods are processed in water (such as the blanching stage that is essential before freezing and drying), when they are canned, or even simply when they are boiled, there is an inevitable loss of part of the water-soluble nutrients. These include the B vitamins, vitamin C, some soluble mineral salts and small amounts (usually of no significance) of proteins and carbohydrates.

Manufacturing losses are often in place of, not in addition to, domestic cooking losses

Many processed foods are already cooked and require, at the most, reheating before being eaten, while foods that have been blanched require a shorter cooking time in the home. Consequently, the loss of water-soluble materials that occurs during the factory process is merely in place of the loss that would occur in the home, as exemplified in Table 3.1.

Table 3.1. Vitamin C lost in frozen and freshly cooked garden peas, as consumed.

Fresh	Frozen	
Boil 6.3 min	Blanch 3 min	loss 11%
(1.1 min to rise to BP)	Boil 3 min	loss 30%
Loss 40%	Total loss	41%

Indeed, there are factory techniques in which such losses are reduced so that total losses are less than those incurred when raw products are cooked in the home.

Losses refer mainly to vitamins C and B_1

Although there is much discussion of nutritional losses in food processing, these involve chiefly vitamins C and B_1. Both are water-soluble and so are partly lost in wet processing; vitamin C is rapidly oxidized by air when cooked foods are kept hot, while vitamin B_1 is sensitive to heat.

Under proper control, other nutrients should be relatively stable, apart from leaching loses. However, the objective of the food manufacturer at the present time is to produce foods that are the most acceptable in terms of flavour, texture and colour—nutrition seems to have little sales appeal—so if processing causes nutritional damage it arouses little interest from either the manufacturer or the consumer.

However, developments in food processing are moving towards gentler methods of treatment in order to conserve flavour, texture and colour; such methods tend also to preserve the nutrients, even where this is not the primary intention. This, added to gradually increasing awareness of nutrition by the consumer should result, in time, in processed foods of higher nutritional quality.

It is essential to take into consideration the role of the food in the diet as a whole

Nutritional damage, even when severe, may be of little, if any, significance in the diet as a whole. Table 3.2 provides a detailed consideration of the role of the food in the diet as a whole. It lists the vitamin content of canned whole meals both before and after processing, as well as after 5 years' storage.

Table 3.2 Losses of vitamins in whole meals during canning and subsequent storage.

Vitamin	Initial value	% loss after canning	% loss during storage 1.5 yrs	3 yrs	5 yrs
A	16.5 μg	50	100	—	—
E	80 mg	0	0	50	50
B_1	9 mg	50	75	75	75
B_2	6 mg	0	0	0	0
B_6	5 mg	0	0	0	0
B_{12}	18 μg	0	0	0	0
Niacin	110 mg	10	20	20	20
Pantothenate	21 mg	25	50	50	50
Folic acid	14 μg	0	0	0	0

The figures suggest considerable losses of vitamin A, but since the recommended daily intake is 750 μg and the raw foods before any processing supplied only 16 μg, this meal would make no useful contribution to vitamin A in the diet, even if there had been no damage to the vitamin.

On the other hand, the thiamin level is so high that even after processing loss and 5 years' subsequent storage (not unusual for canned foods) the meal still supplies about twice as much as the recommended daily intake of thiamin.

The table also shows that many vitamins are completely stable after canning and storage.

Special consideration must be given to the vulnerable groups of society

The average adult eats so varied a diet (in communities where food, and money to buy it, are available) that severe damage even to a major food would have little noticeable effect on his nutritional status. People living on very restricted diets, however, rely so heavily on a limited number of foods that they may be at risk from excessive nutritional damage. The most important vulnerable group are those infants who rely entirely on a milk preparation from one manufacturer. Indeed, it was just this problem which, as late as 1954, demonstrated that vitamin B_6 was essential for human beings; previously it had been shown to be essential only in experimental animals. A minor change in processing, involving a higher temperature, so reduced the amount of vitamin B_6 that a number of cases of convulsions resulted. Such a product would not, of course, have had any effect on adults or older children who receive adequate amounts of this vitamin from many other sources. An analogous problem arose with potatoes. The potato is not a particularly rich source of vitamin C but, because of the amounts eaten, it provides about one-third of the average intake of vitamin C for a large part of the population of Western Europe. In Great Britain in winter, when many fruits and vegetables are less plentiful and consequently more costly, potatoes provide as much as one-half of the average intake of vitamin C, so it is a food of vital importance to a considerable number of people.

The potato also supplies about 15 per cent of the intake of thiamin. Both these vitamins can be considerably damaged in processing—in this case damage additional to that taking place in domestic cooking.

In the early 1960s, instant dried potato was reintroduced onto the British market (after a brief wartime introduction) and the vitamin C content of some brands of this product was very much lower than that of the unprocessed product. Since the replacement of the traditional uncooked potato by the instant variety was only partial, and also slow, this did not result in any nutritional problems, but the possibilities are clear, and the lack of consideration given to this problem by manufacturers was regrettable, since the instability of vitamin C and our dependence on the potato were well known.

Ready-peeled and ready-chipped potatoes are treated with sulphite to prevent browning. Sulphite damages thiamin, so again, although no nutritional effects have come to light, this poses a potential problem.

These examples serve as a warning, to both manufacturers and public health authorities, of the continuing need for monitoring food supplies and their possible effects on health.

Advantages of processing

Much discussion concerns nutritional damage, but there are many benefits. The most obvious benefit is that of preserving foods (by sterilization, pasteurization, drying, chilling, freezing and the use of chemical preservatives and antioxidants). Before the advent of factory processing, as still today in many primitive communities, people relied largely on the previous crop (and on foods they had preserved and stored for themselves) and shortages occurred between harvests. Climatic conditions led to frequent famines, even in Western countries, before the present century.

It is not always realized that fresh foods lose some of their nutrients as soon as they are harvested, especially, present research tells us, vitamin C and folic acid. This loss is due to the presence in the food of enzyme systems which destroy the vitamins when the cell walls of the plants begin to break down, as the leaves wilt or the food is bruised in handling. Under warm conditions that favour wilting, as much as half the vitamin C and folic acid can be lost from leafy vegetables in a day. Consequently, processing maintains the levels of these vitamins after the initial loss caused by the process itself.

The simple domestic process of chopping or mincing, or even slicing salad foods, can result in almost complete destruction of vitamin C before the food is consumed, depending on the time between the 'damage' and consumption. Certainly, some salad foods can lose half their vitamin C in 2-3 hours.

Furthermore, the manufacturer has access to fresher foods than the housewife. The term 'fresh' is usually applied only to raw foods, and it is possible to have foods that are garden fresh (meaning just harvested) or market fresh, i.e. the food that the housewife buys in the raw (but possibly stale) condition. The manufacturer is able to start his process with foods richer (at least in vitamin C and folate) than is the housewife, so that he may finish with processed foods richer in these nutrients than the home-prepared product—even assuming that the housewife is competent in this area.

There are also direct nutritional improvements from processing. The greater part of the niacin of cereals is present in a chemically bound form which is not biologically available and this is largely liberated during the process of baking, especially if alkaline baking powders are used.

The niacin in coffee is produced during the process of roasting. The very process of cooking renders more of the vitamins available, since it can be shown in the laboratory that many of the vitamins cannot be extracted completely from some foods even if homogenized, unless the cell walls have been broken down by heat. Consequently, the supposed superiority of raw fresh foods over cooked and processed foods may be quite untrue in some cases. However, this last statement is an extrapolation from laboratory data, and no attempts have been made to ascertain whether human beings extract more or less nutrients from raw rather than processed foods.

A further advantage of processing is in the destruction of toxins. This applies particularly to legumes, some of which are toxic when raw or incompletely cooked. (Although the toxins are present in other raw foods, their concentration is so low that they normally have no ill-effects.)

Raw kidney beans of the variety *Phaseolus vulgaris* contain substances that cause agglutination of the red blood cells, and in Great Britain between 1975 and 1980 there were some 100 outbreaks involving over 800 individuals from eating raw or incompletely cooked beans. This does not occur with processed (i.e. canned) beans, since the sterilization process that is the essential part of canning completely destroys all toxins.

The whole area of food consumption, whether the foods are raw or processed, chemically treated or otherwise, is a balance between risk and benefit. Many raw foods contain known toxins; even if they are present in small amounts it would, presumably, be safer to avoid such foods, but this would limit the variety of the diet and might have an adverse effect on nutritional status. Similarly, the process of cooking, especially frying, produces small amounts of known toxins. This means that when the substances formed are extracted and fed to experimental animals in abnormally large amounts they cause changes which can be fatal. Obviously, while these toxins, natural or produced by cooking, are present in amounts small enough to be dealt with by the body, their effects are certainly not cumulative. However, animal experimentation is the very method used to test the safety of manufacturers' food additives. They are fed in extremely large doses continually to experimental animals, and unless they are shown not to cause harm under these conditions, their use is not permitted.

The problem is complicated by the finding that almost everything—including vitamins and water—is toxic, and indeed can be fatal if consumed in excessive amounts. Food additives are usually investigated to determine the maximum amount that an animal can tolerate *without* any harm, and then to permit *one-hundredth* of this to be used in the diet. Leaving aside at the moment the problem of food additives (discussed later) and concentrating on the relative advantages and disadvantages of food processing, there are varying ratios of benefit and nutritional risk. Since these processes were introduced before we had gained our present knowledge of nutrition, no manufacturer or legislative authority was called upon to make a judgement, but with the development of novel foods and novel processes these problems are now arising.

Current experience provides examples of varying ratios of benefit and nutritional risk.

(1) The pasteurization of milk results in the destruction of about 10 per cent of the thiamin and riboflavin and about 25 per cent of the vitamin C, and smaller losses of other nutrients. Most consumers regard such losses as a price worth paying for safe milk.
(2) In Great Britain comminuted meat products (such as sausages) are preserved with sulphite. Ignoring any advantages of hygiene, such preservation means that the manufacturer needs to deliver to retail outlets less frequently, so (in theory at least) the product is slightly cheaper, but devoid of the thiamin. Is this a price worth paying for lower cost?
(3) It has become convenient (and cheaper for caterers) to buy potatoes ready-peeled and ready-chipped. They must be protected from browning by treatment with sulphite solution which, as mentioned above, damages some of the thiamin. Is this a price worth paying for convenience?

One answer would appear to be that unless, or until, some health risk becomes apparent, then the prices paid for such advantages are acceptable.

Measurements must be made of the food 'on the plate'

There are many reports in the literature of the changes taking place during processing, but it is the food eaten that provides the only relevant results. Table

3.3 shows that fresh garden peas cooked under ideal conditions inevitably lose 56 per cent of their vitamin C. If this had not been included, it could have been deduced from the table that processing causes considerable destruction of this vitamin. When the foods are compared 'on the plate', however, it can be seen that there is little difference between freshly cooked peas and those processed by freezing and freeze drying. Even peas subjected to the more severe processes of canning and air drying are not so very different from the freshly prepared product. Indeed, if we only knew how much damage is being done to nutrients in the home, an area where little is known, we might find that some processed foods are nutritionally superior to home-prepared ones.

Table 3.3 Nutrient content of 'food on the plate'.
Garden peas: percentage loss of vitamin C after stages of processing.

Fresh		Frozen		Canned		Air dried		Freeze dried	
		Blanching	25	Blanching	30	Blanching	25	Blanching	25
		Freezing	25	Canning	37	Drying	55	Drying	30
		Thawing	29						
Cooking	56	Cooking	61	Heating	64	Cooking	75	Cooking	65

Factory v. Home Preparations

In some instances direct comparisons are not possible because the factory foods—butter, margarine, cheese, dried milk powder, breakfast cereals—cannot be prepared at home. In other instances the housewife uses more expensive ingredients, or larger amounts of major ingredients, such as meat, fish or fruit, than is feasible in mass-produced foods if they are to be sold cheaply. While this does not mean that they are necessarily higher in nutritional value or better in eating quality, it is likely that they will be. On similar lines, the housewife may use eggs in mayonnaise and in cakes, full-cream milk in milk puddings and lemons in lemonade, while chemical substitutes and cheaper fillers are used in the factory.

Finally, after all these considerations, in many instances the choice is not between processed and home-prepared foods, but between processed foods and none at all. An often quoted example is that of the garden pea, which is harvest-fresh in Great Britain for only some 6 weeks. Apart from this short period in which the fresh product is available, the choice would tend to be between processed (i.e. frozen, dried or canned) peas and none at all.

LOOKING TO THE FUTURE

Looking to the future, two issues are of particular interest. One is the direction of possible nutrition research. The other is the likely food supply of the future (Bender, 1980; NAS, 1974).

Research Directions

Individual variation

Reference has been made to recommended intakes of nutrients which would satisfy the needs of population groups, and to dietary goals applicable to communities. However, we lack information on the needs of individuals.

People appear to be able to adapt to intakes of nutrients well below those levels recommended for population groups as a whole. Although there is a continuous destruction of nutrients in the body (which explains why there is a continuing need for a regular intake), the body adapts to low intakes by reducing the rate of destruction. It is not known how far this can go, but it is certain that many people, in affluent as well as in developing countries, are consuming much less of some nutrients than is, theoretically, essential. Whether the state of health of an adapted person is as good as (or perhaps even better than) one who is consuming greater amounts is unknown.

It might be possible in future to monitor the health of an individual and detect any departure from normality before damage results. If his various blood and tissue parameters were measured from an early age when he was in apparently good health, and at regular intervals thereafter, this would establish his 'normal' range of blood constants, as distinct from group averages. Any deviation would serve as an early warning of problems, which might then be avoided.

Identifying those at risk

There is known to be a genetic component in disorders such as coronary heart disease, obesity, diabetes and, indeed, in the length of life. If those at risk could be identified, it might be possible to modify the progress of the disease through dietary manipulation. Three diverse examples illustrate the possibilities.

Obese people appear to be born with the tendency to become fat if their energy intake exceeds expenditure, in contrast to those who simply burn off any surplus. If they could be identified early, it might be possible to prevent obesity developing.

The second example is that of blood platelets, which play an important role in blood coagulation, and consequently in thrombosis. It appears that the platelets produce an identifiable protein in the blood, and if the amount is found to be related to subsequent thrombosis it would be possible to increase the stability of the platelets through the diet (less saturated fat, additional calcium and even the ingestion of alcohol).

Hypertension is the third example. This is common in all Western countries and a complicating factor in a range of disorders. Some 15 per cent of the population appear to be genetically susceptible to a high sodium intake, and would benefit from a reduction.

Specific advice to susceptible individuals is far more likely to be accepted and successful than general advice to the public as a whole. The problem is to identify these people.

Food intolerance

Individual variation applies particularly to food intolerance. This is the term applied both to true allergy (which involves antibody formation) and to personal idiosyncrasies, which do not involve antibodies.

The potential of related research is illustrated by the successful dietary control of a range of genetic diseases that have come to light over the years. Most of these 'inborn errors of metabolism' are rare and affect fewer than one hundred people in a million, but it is now considered worth screening babies at birth.

About one-third of the population suffer some degree of food intolerance, and have learned to avoid the foods responsible, but evidence suggests that there are many hidden cases. When the effects of an allergy take 48 hours to show, it is unlikely that any foodstuff will be suspected. Further research in this area is likely to yield considerable benefit.

Environmental toxins

Many ordinary foods contain natural toxins. The popular belief that so-called natural foods, usually meaning those consumed without being subjected to any method of processing in a factory, are 'safe', in comparison with processed foods, is quite incorrect. Indeed, the opposite is often true, since processed foods are subjected to public health regulations, and food ingredients are tested for safety, while 'natural' fruits and vegetables are not.

Most, if not all, fruits and vegetables contain toxic substances. That is to say, when they are extracted and fed to experimental animals, their effects are such that, if they were food additives, they would not be permitted in the diet. Potatoes contain the toxic alkaloid solanine, rhubarb contains oxalic acid, many legumes contain cyanide, and there are many thousand such chemicals in our raw foods.

They are, of course, present in very small amounts; the body can detoxicate and eliminate such compounds, otherwise presumably mankind would have died out. On the other hand, we do not know how much ill-health is caused by such natural toxins, or whether they worsen infections and other problems.

Cooked foods also present a problem. Proteins can form lysinoalanine when heated. This substance, when extracted from heated foods and fed to experimental rats, causes kidney damage. Since most foods contain proteins even in small amounts, this means that foods other than purified fats and starches can form this toxic agent when heated. No harmful effects have been found in human beings from this cause, but we cannot be certain that the regular consumption of lysinoalanine, or heated protein-containing food, has no long-term effect.

Interrelation between nutrients

A great deal is known of the metabolism of individual nutrients but little of the interrelation between them.

It has been established that some enzyme systems include vitamins and minerals, such as zinc or copper, or are activated by these ions, while the metabolism of protein requires vitamin B_6 and the utilization and function of the ubiquitous ATP (adenosine triphosphate) require magnesium. In magnesium deficiency, the sodium-potassium balance is disturbed as a result of disturbed magnesium and calcium balance in cell membranes. There is evidence that the effect on susceptible individuals of too high an intake of sodium in relation to hypertension may be modified by an increased intake of potassium and that increased consumption of polyunsaturated fatty acids—which reduce the levels of blood cholesterol—calls for an increased intake of vitamin E.

These are relatively simple examples of the type of interactions that occur and about which little is known.

Almost all dietary changes result, in the short term, in marked departures from 'normality'. For example, an increase in the intake of wholegrain cereals, which include dietary fibre and phytic acid, results in reduced absorption of zinc, iron and calcium. Reduced levels of zinc in the blood plasma have been reported in such diets. But since many communities live in apparently good health on such diets and cannot possibly suffer continuously reduced absorption of these nutrients, the body must adapt after the initial change.

This is an important area for future investigation. Fortunately for mankind, our diet is usually mixed, meaning that we consume different foods from day to day. This is in marked contrast to experimental animals which are fed the same diet for periods of time which are equivalent to many months or even years in man. While such control is essential in experimental work it means that we have, as yet, little information on the effects of any dietary changes in free-living human beings.

Furthermore, we know very little of the long-term effects of diet. For example, the loss of bone tissue (osteoporosis) is an inevitable concomitant of ageing, especially in women, and leads to fragile bones which break easily. It is not known whether a denser bone produced in early life by a diet richer in calcium and vitamin D would mitigate this—and short of a life-long controlled experiment it would be difficult to verify.

Clearly, such research is far more complex than the earlier investigations into the 'simple' deficiency diseases.

Food Supplies in the Future

Food surpluses, such as those accumulated in the European Community, and acute famines which seem to occur so frequently in the poorer countries, are short-term affairs. In the long term, we have to view actual and potential food supplies against the rapidly growing population of the world. It is calculated that there is a 10-20 per cent shortfall at the present time; consequently, production will have to increase by 120-140 per cent over the next 20 years to feed the doubled population. The question is, then, 'What will our children eat?'

One answer is, 'Exactly what is eaten now, although produced in different ways.' Selective breeding of crops and animals can provide high yields; the use of more economic substitutes and the conversion of waste into edible foods must be aimed largely at providing the same types of food that people are eating now—otherwise it is unlikely to be accepted.

Considerable effort has been devoted to producing novel foods. Much of this effort, it is now realized, was devoted towards solving the non-existent 'protein problem'. It was thought in the 1960s that while energy foods (mostly starch) could be produced in adequate quantities, there was a shortfall in protein foods. Now, it is known that where there are shortages, it is *food as such*, rather than protein, that is in short supply. The major foods of the world are the cereals, and they contain adequate amounts of protein. In other words, so long as there is enough food to satisfy energy needs, then there will automatically be enough protein in the diet (although the needs of weanlings present a special problem).

For convenience in discussion, novel foods may be divided into five groups, namely synthetic foods, substitute foods, primary foodstuffs, food from waste and 'single cell protein'.

Synthetic foods

The basic foodstuffs, such as fats, carbohydrates and polymers of amino acids, have been synthesized but cannot be compared, in terms of cost and scale, with farming. Nor is it at all likely that they will play any part in the future.

Two amino-acids, namely lysine and methionine, are available at acceptable prices and are used to a limited extent to improve the quality of the protein of animal diets, with consequent reduction in quantity of protein. In view of the earlier statement that cereals supply enough protein, however, there is no purpose in enriching human foods with these amino acids.

The only foodstuffs that it is useful to synthesize at the moment are the vitamins.

Substitute foods

The second group, substitute foods, is epitomized by textured vegetable protein (TVP), involving the use of raw materials such as ground nuts, field beans and particularly soya bean to make a range of foods. Although the simulation is not perfect, great technical success has been achieved by food technologists in spinning, extruding, texturizing and flavouring, and it is possible to produce them at relatively low cost on a large scale. Such processing combines the crop efficiency of vegetable foods with the taste of meat.

Another widely known substitute food is coffee whitener. Technical skill has achieved the whitening power, diffusibility and flavour required in a white powder that is stable for much longer than the liquid milk that it replaces.

A different type of substitute is food such as 'synthetic' rice, granules made from groundnut meal and tapioca flour which are, or can be at the relevant time, more freely available than rice, and could be made with improved nutritive value.

Primary foodstuffs

The third group of novel foods are the primary foods, i.e. those that could be consumed directly instead of passing through the biological chain, with consequent loss of food value. One example is plankton, on which the fish that are ultimately consumed basically rely. It is calculated that it takes 10 000 tonnes of plankton to yield 1 tonne of edible fish so, in principle, food supplies could be increased by this proportion. Plankton can be collected by dredging and they are reported to have the taste of anchovies. However, long before energy reached its present price level, it was generally agreed that the cost of such collection far outweighed the value of the food. Another and more likely primary food is krill, the shrimp *Euphausia superba*, which provides food for certain varieties of whales. These are about 2.5 cm in length and the potential annual catch is said to be in the order of 50 to 100 million tonnes a year. The krill are found in high concentration of several kilograms per cubic metre of sea in swarms up to several hundred metres across and going down to great depths. At the present time an Australian research group is examining the possibilities in the Antarctic with a view to the management and conservation of this foodstuff. As well as the whales, a number of species of fish and birds rely on it for food, so it needs to be conserved. In developing its use as a food for humans, considerable technical success in removing the shell mechanically has been achieved in Chile. Thus, it appears that this is a feasible long-term project.

Leaf protein serves as an excellent example of most of the factors involved in the development of novel foods. First, it is not new, having been first suggested in 1773. Then there is the fact that the procedure involved is vastly more efficient than traditional crop production, since leaf protein can be obtained by repeated cropping rather than waiting for one or even two main crops each year. Cropping green wheat leaves repeatedly yields 25-30 times as much protein as can be obtained from the same crop as cereal grain. The actual efficiency is much higher than this indicates, because ordinary food plants do not make maximum use of the energy of the sun that is available, whereas different and far more efficient plants can be grown for the purpose of extracting the protein.

When forage crops are fed to animals for conversion into human food the efficiency is 10-30 per cent, but 40-60 per cent can be extracted for direct use by man. As with all these 'new' developments with phenomenal promise, the practical application has been disappointing. After some 30 years of development, the main use of leaf protein is in feed for poultry and pigs. (See also Table 3.4.)

Table 3.4 Relative efficiency of production from plant and animal sources.

Animal sources		Plant sources	
Millions of kilocalories per hectare per year			
Beef	0.4	Rice	6.6
Eggs	0.5	Potatoes	11.6
Milk	1.8	Bananas	12.6
		Sugar	25.0
Kilograms of protein per hectare per year			
Dairy cows	10	Potatoes	29
Chickens	6	Cereals	33
Beef	4	Grass	80
Pigs	4	Beans	80

Food from waste

Fuel costs, the state of the world economy and consideration for the environment have led to increased interest in making use of waste. This is not new in the food field, where the re-use of organic material has, over the years, been governed largely by price. Whey from cheese making is poured down the drain; although it contains protein, carbohydrate, minerals and vitamins, the cost of its recovery is not worthwhile. The same has been true, at different times, of blood from the slaughterhouse, protein residue from potato and corn starch manufacture, and other by-products. In Malaysia waste from canning pineapples is being used as the substrate for yeast intended for animal feed; in Indonesia the liquid left after preparing soya curd is used as a substrate for growing edible mould, Neurospora; in Brazil mushrooms are grown on molasses. The effect of varying economic conditions is illustrated by the conversion of sawdust from timber mills into glucose and thence into alcohol—a process used early in this century but outdated by the introduction first of cheap molasses and later of synthetic alcohol.

Leaf protein can be extracted from waste crops such as cotton, banana, jute, sugar-beet and peas, so it can also be included in this group.

Single cell protein

The work on single cell proteins, i.e. bacteria, yeasts, algae and also moulds, is well known, and some of the results have already been put into practice. The deep interest can be gathered from the figures for production. So far as land space occupied is concerned, these organisms are extremely efficient relative to conventional plants. One hectare of land will produce 10 kg of protein as meat, 17 kg as milk, 80 kg as peanuts, 110 kg as grass and 2600 kg as algae. The micro-organisms fall into two groups: those which require to be supplied with a source of energy for growth and reproduction (carbohydrate or hydrocarbons) and those which use sunlight as their source of energy, namely the algae. These latter need to be supplied only with carbon dioxide, mineral salts and a source of available inorganic nitrogen such as ammonium sulphate. The current favourite is the blue-green alga, *Spirulina*, which fixes its own nitrogen from the atmosphere and so uses energy even more economically. It was discovered (at least by modern science) as recently as 1962 in North Africa, near Lake Chad, where it is consumed by the locals, and also in Mexico, where it had provided food for the Aztecs centuries earlier. The potential for *Spirulina* is shown by the yield calculated on an annual basis of 40 tonnes of dry matter per hectare, under the best conditions of cultivation; yields of 10 tonnes have been obtained under what are termed semi-natural conditions.

Yeast has been developed over the past 25 or 30 years and has reached factory-scale production intended for animal feed. The energy source has been by-products from the petroleum industry, and its successor, the bacterium, is grown on methanol derived from natural gas. Here again, the potential is gigantic—it is calculated that in the Gulf oil production areas some 50 to 100 million tonnes of gas are 'flared off' annually.

Old foods

Some 3000 species of plants have been used for human food at different times and in different countries, but only 150 of these have been cultivated commercially. In fact, we rely for the greater part of our food supply on only 20 crops. There are many others that are well known but used only in a very limited locality. They show promise for commercial production. If these achieve their potential, then they will make some impact on the foods that we, or at least our children, will eat.

The carob or locust bean (*Ceratonia siliqua*) is grown in the subtropics. In Cyprus it is grown for its gum, which contains up to 60 per cent protein; the bean itself contains 21 per cent protein.

Two species of the Chenopods, the quinoa and the canihua (*C. quinoa* and *C. pallidicaule*), are gathered for food throughout the Americas, but cultivated as a major item of the diet only in the Altiplano region of the Andes, and are unknown in Europe. They tolerate light frosts and limited rainfall, and can grow on soils where other cereals grow poorly. Quinoa contains 12 per cent protein and canihua 14 per cent ; up to 20 per cent can be mixed with wheat flour for bread making.

A number of plant scientists were polled for their opinion regarding promising food crops not currently exploited and 400 were suggested. In 1974 a meeting of experts shortlisted 36 of these as being most worthy of further examination (NAS, 1974). Some of these show considerable promise. The grain amaranth has long been grown in South America but it is now used as food only in a very limited locality. Grown in its wild state, the yield is greater than that of maize. When improved strains are developed and fertilizers and irrigation are applied, the potential yield would appear to be considerable. The grain develops on seed heads like sorghum and contains 15 per cent protein, which is higher in quality than the commonly used cereals.

There are many more examples of these foods, which may be novel to us but are common in certain areas of the world. Apart from these developments, which require lengthy selective breeding trials, there are foods which have already spread from their place of origin across the world. Avocado pears and kiwi fruit provide two examples where food science and modern transport have introduced new foods to the Western European countries. Such innovations will undoubtedly increase, but the basic foods that people prefer will, in my view, remain largely as they are now, despite the nutritionist and the food technologist. Until we can ascertain why certain foods such as the hamburger and the potato chip have become universally and apparently permanently popular this will continue to be true—but this is a question which lies outside the scope of the present chapter.

REFERENCES

Anon (1980) 'Why the American decline in coronary heart disease?' *Lancet,* **i**, 183-184.

Bender, Arnold E. (1978) *Food Processing and Nutrition*, Academic Press, London, New York.

Bender, Arnold E. (1980) 'What will our children eat?' *Chemistry and Industry*, 18 October, 797-800.

DHSS (Department of Health and Social Security) (1978) *Prevention and Health—Eating for Health*. HMSO, London.

NACNE (National Advisory Committee on Nutrition Education) (1983) 'Proposals for nutritional guidelines for health education in Britain'. Health Educational Council, London.

NAS (National Academy of Science) (1974) 'Underexploited tropical plants with promising economic value'. NAS Washington, DC.

Part II
Why We Eat

The Food Consumer
Edited by C. Ritson, L. Gofton and J. McKenzie

CHAPTER 4

Economic Factors Influencing Food Choice

STEFAN TANGERMANN

WHO BELIEVES IN ECONOMICS?

Economists are funny people. They try to explain—and love to prescribe—human behaviour. You would expect that in order to do so they would be intensively engaged in studying the way in which people actually behave. It would seem natural to believe that economists would go out and observe what individuals do and ask them why they do what they do. Nobody would be surprised if economics turned out to be, essentially, a special application of psychology and sociology. Only economists would be surprised if you suggested that this should be the case, as economics—at least traditional mainstream economics—is a completely different type of science. New students of economics, who expect their professors to tell them how real people behave in economic affairs, are often deeply frustrated when they discover that they are expected to forget about many facets of the real world and to concentrate, instead, on an abstract model, on an unbelievable caricature of man, whom the economists call homo economicus.

This synthetic person, which appears in all basic economics textbooks, is completely rational and pursues one clear-cut objective, to maximize its 'utility'. It quickly gathers all relevant information, spins it through an exact mathematical formula and derives an impeccable decision as to what to buy or to produce. This miserable creature would not survive one day in real life, lacking all abilities which are required for a successful struggle for existence and any charm which would make it acceptable to human beings. But it happily strolls about in the heads of economists and is used by them as a starting point when they try to explain what goes on in the economy.

If this is true, how, then, can economics be of any use in understanding real phenomena? In our particular context, is it worthwhile to consider economic factors influencing the demand for food, if an analysis of these factors is essentially based on such a distorted view of human behaviour? Should the reader not quickly turn to the chapters on anthropological, sociological and psychological issues? Hopefully he or she will not, but rather will try to grasp what economics has to contribute to our understanding of food consumption. After all, food consumption is very much an economic activity too. A major share of most households' incomes is spent on the purchase of food, and the amount and the type of food bought very much depends on economic factors. Thus, there is good reason to be concerned with the economics of the demand for food. But it is essential to appreciate right from the beginning what economic analysis can, and what it cannot, contribute to our understanding of food consumption and, hence, what will be included in a discussion of the economic factors influencing the demand for food.

Though economic analysis apparently deals with a certain fraction of human behaviour, it is not essentially orientated towards the behaviour of individual persons but, rather, towards the aggregate outcome of activities of groups of persons (food consumers) or the behaviour of an 'average' or typical person ('the' food consumer). Given the complexity of individual behaviour, it would be extremely difficult to aggregate over concrete individuals and still arrive at a clear picture. Hence economists have tended to abstract from most details of human behaviour, to concentrate on what *in economic terms* are the most prominent elements of behaviour, and thereby they have provided a basis for understanding and, to a certain extent, predicting the aggregate outcome of individual actions in groups of people. The results of such an exercise, and therefore the following description of the economic factors influencing the demand for food, are thought neither to give a complete picture of all factors behind the activities under discussion (though it may sometimes appear as if the economic analysis tries to explain everything) nor to provide a realistic impression of any one individual's behaviour (though it may occasionally sound as if this were directly related to individuals). From the manifold issues involved in phenomena like food consumption, economic analysis deliberately deals with only a very limited subset and it does this by way of generalizing rather than by looking into particular actions of people. Given this modest programme of economic analysis, it is reassuring to note that empirical economic analysis has very often been able, as we saw in chapter 2 and as discussed further below, to explain quantitatively a significant proportion of changes in the demand for food.

It may help to illustrate the point with an example from physics. For a long time physicists were not able to say very much about the actual behaviour of the individual particles within atoms. Rather, they worked on the basis of an idea, of a model of what happens within an atom. However, this model was very successful in explaining and predicting most dimensions of the behaviour of the aggregate—*the* atom. Economics, which still is a relatively young discipline if compared with physics, may be thought of as being at about this stage of its development. There are many attempts at increasing the degree of complexity and thereby reducing the degree of abstraction with which economics operates. The time may come when the homo economicus carries more flesh on his abstract skeleton.

Having made all of these remarks on the nature of economic analysis, we have still not discussed what may appear a more immediate question, namely what are *economic factors*, as distinct from other types of issues behind food consumption. An unprejudiced person may think that 'economics' is everything which has to do with money. Economists would strongly oppose this statement and explain that they deal with many issues which are not at all related to money. Asked to define what economics is, they are, however, somewhat helpless and tend to answer with the old joke that 'economics is what economists do'. In our context, however, things are relatively easy. When we talk about economic factors influencing the demand for food we essentially have in mind incomes and prices. To the extent that we are concerned with industrialized countries, the majority of consumer incomes are in money form and prices are usually paid in money rather than in kind. Hence in our context, the naïve definition actually applies—when we talk about economic factors behind food consumption we have those factors in mind which are related to money. In particular, we are interested in the way in which the quantities and types of food bought by consumers react to changes in prices and consumer incomes. This is a very limited programme, but there is quite a bit to say about it.

MR ENGEL AND THE FOOD CONSUMER

One of the relationships we are interested in, that between food demand and income, belongs to those economic phenomena on which very much empirical research has been done. Indeed, it was the subject of one of the most famous early investigations in the area of empirical economics. In 1857 the Prussian statistician Ernst Engel published the results of a study in which he looked into the expenditure patterns of families with different levels of incomes. As far as expenditure on food was concerned, he found out that 'the poorer a family is, the greater the proportion of total expenditure which it must use to procure food. The wealthier a people, the smaller is the share of expenditure on food in total expenditure' (Burk, 1962). This statement appears completely plausible. Rich people can afford to buy more food than poor people. However, there is also the phenomenon of increasing saturation with increasing food consumption. Hence the willingness and, indeed, ability to eat food does not grow in proportion to income.

Engel's observations have very often been repeated in many countries and at different times (David Blandford's OECD study (chapter 2) provides a recent example). Table 4.1 presents data for postwar Germany which exhibit the same basic pattern. The observation that the share of food in total household expenditure decreases with increasing income has, indeed, been so widespread that it has become usual to refer to 'Engel's law'. Of course, this is *not* a law in the sense of laws in the natural sciences, as there is no absolute reason for it to apply universally. In economics, as in all social sciences, there are no 'real' laws in that sense. However, Engel's 'law' (or rather this generalized observation) has been found to hold in a great many situations.

Table 4.1 The share of food in total consumer expenditure at different income levels, West Germany.

Income class	% share of food[a] in total expenditure		
	1965	1972	1980
Low income	50.2	42.2	33.5
Middle income	40.0	33.3	28.1
Higher income	28.5	25.5	22.0

Note: (a) Including beverages and tobacco as well as consumption in restaurants.
Source: *Statistisches Jahrbuch uber Ernährung, Landwirtschaft und Forsten*, various issues.

To clarify this point about the law-like nature of Engel's observation on the relationship between food expenditure and income, let us consider a very poor family, say in a developing country. Out of an annual income of $400 it may spend as much as 70 per cent, i.e. $280, on food and still be malnourished. Should its income grow to $440 it may use the whole increment of $40 to buy more food, thereby increasing its food expenditure to $320. The new share of food expenditure in income is then 72.7 per cent. Thus, in this particular case, the proportion of food expenditure to income has increased rather than decreased with increasing income. Hence in this situation, which is not at all unrealistic, Engel's 'law' does not hold.

In considering an economic relationship like that represented by Engel's law, we have implicitly done something which economists do very often. We have analysed the isolated relationship between only two variables, in our case income (or total expenditure) and expenditure on food. In doing so we have implicitly assumed that all other factors which could determine food expenditure do not matter. Alternatively, as we of course know that there are many relevant factors apart from income, we have assumed that all these other factors remain constant. This is the famous (or infamous) *ceteris paribus* clause in economics. Natural scientists can make experiments in which they have complete control over the whole environment. They can then change one factor at a time and observe the results. They have constructed a *ceteris paribus* case. In the social sciences, for the most part, such experiments are not possible. Of course we could make an experiment and furnish a given family with 10 per cent more income for a certain period, but this can never be the true *ceteris paribus* experiment of the natural sciences, since we cannot replicate the factors in the original situation. In the second period, not only will income be different, but other factors which we cannot control (such as age of family members, taste, neighbours' consumption, etc.) will also have changed. Thus, we can make a 'thought experiment' only, *assuming* that income alone has changed and asking ourselves to what this could lead.

When we want to confront this intellectual experiment with reality, in order to find out if our hypothesis is acceptable and which magnitudes are involved, we are in great difficulty, as we have to extract from the total change in the variable in which we are interested (say food expenditure) the effects of all other factors (say taste changes, etc.) before we can determine the influence of the factor we want to analyse (say income). We shall have to return to these difficulties of empirical economics below, but let us apply some of these considerations to Engel's law here.

Engel, too, necessarily had difficulties with the *ceteris paribus* problem in his empirical research. In a cross sectional analysis, he compared different families with differing incomes at one point in time. Of course, their differing food expenditures were not due only to differences in income. Thus he should ideally have tried to eliminate the influence of all other factors before he measured the influence of income. He was able to do this for at least one important non-income factor, i.e. family size. The statistics available to him were such that he could group families according to size in order to compare food expenditures among families of the same size but with differing incomes. At the same time, by comparing families with equal income levels but different sizes, he was able to conclude that the share of food in total expenditure tends to be greater, the larger the family. Of course, this is fully consistent with his observations regarding the influence of income as, with given family income, the larger the family, the less the income per head.

There are many other factors which influence food expenditure. An important group with which economists usually do not deal very much is labelled 'taste', 'preference' or 'consumption habits' in economics. While the influence of this group of factors is very pronounced at the level of individual food items, it is effective with regard to overall food expenditure as well. Table 4.2 provides a comparison among member countries of the European Community. It is interesting to note that Engel's law roughly applies to a cross-section of country aggregates as well. However, it is somewhat 'spoiled' by the influence of differences in consumption habits among countries.

So far we have dealt only with aggregate food expenditure. What can we say about the relationship between income and the quantities of individual food items consumed. Can we directly translate Engel's law to the quantities of individual foods? Here we have to be cautious. Apart from possible (and likely) difference between individual types of food, to which we shall soon come, we have to note that there are two important differences between the development of food expenditure on the one hand and the quantity of food consumed on the other. First, even individual food categories, say fruit, consist of different

Table 4.2 The share of food in total private expenditure in member countries of the European Community, 1980.

Country[a]	Gross Domestic Product per head, $	% share of food[b] in total private expenditure
West Germany	13269	21.3
Denmark	12961	25.8
France	12184	21.4
Netherlands	11974	19.6
Belgium/Luxembourg	11927	21.8
United Kingdom	9368	21.9
Italy	6956	30.7
Ireland	5280	44.3[c]
Greece	4164	41.6

Notes: (a) Countries are arranged in descending order of income.
(b) Including beverages and tobacco.
(c) 1979.

Source: EUROSTAT (1983), *Yearbook of Agricultural Statistics.*

commodities—apples, pears, citrus fruit, etc., with different prices per kilogram. It is likely that the composition of a food category changes with changing income. Usually we observe that, with increasing incomes, consumers switch to more luxurious, i.e. more expensive, food commodities, say from apples to citrus fruit. To the extent that this occurs, with increasing income, quantities consumed increase less than expenditure. Second, consumer prices and, therefore, consumer expenditure include, besides the value of the raw food commodity, a margin for processing and distribution—sometimes known as the 'marketing margin'. For reasons which will be discussed below, this margin tends to increase with increasing income. Hence, for this reason, too, quantities of food consumed tend to increase less than expenditure when incomes grow.

These two aspects taken together mean that, if Engel's law holds for food expenditure, it holds even more for the quantity of food consumed in aggregate. We can, therefore, say that with increasing incomes the quantity of food consumed generally increases less than proportionally. Of course, there are considerable differences between different types of food. As we have already argued, consumers tend to switch to higher valued commodities when incomes increase. Expressed differently, this means that consumption of higher valued commodities reacts more strongly to income growth than consumption of more basic foodstuffs. What is considered a higher valued commodity differs from country to country and from period to period. However, items like meat, cheese, fresh dairy products, fruit and beverages would, in many cases, fall into this category. Though in most developed countries consumption of these commodities grows less than proportionally when incomes rise, it is in most cases still significantly and positively affected by income growth. (Blandford's study underlines this conclusion in the case of animal products.)

At the other end of the scale are those commodities the consumption of which is negatively correlated with income. In the jargon of economics they have come to be labelled 'inferior goods'. In developed countries they tend to be calorie-rich basic foodstuffs like potatoes, bread, and in some cases also sugar and butter. At low income levels these relatively cheap commodities are consumed in order to meet basic calorie requirements, while with rising incomes consumers can afford to switch away from them towards more attractive but also more expensive types of food.

Figure 4.1 provides some insight into these relationships and also into the significant differences between the food consumption patterns of selected countries. Quantities consumed per head are plotted against aggregate income per head (measured as national Gross Domestic Product, GDP). Each point in this graph relates to an individual year for the country concerned; therefore the graph shows how consumption and income have moved together through time. It is important to note that this implies that the *ceteris paribus* clause is not valid in this case. Thus, for example, if the graph appears to suggest that butter is an inferior good in the United States because butter consumption has decreased while incomes have grown, this is not *necessarily* the case. It may also have been that butter consumption went down *in spite* of growing incomes because butter prices have risen, consumer preferences have switched away from butter, etc.[1]

However, with the necessary caution, some tentative conclusions can be drawn from this graph. Meat and cheese consumption (and, as an input into animal production, utilization of feed grain) appears to be positively correlated with income in all countries covered.[2] Grain for direct human consumption (in the form of bread, etc.) exhibits properties of an inferior good. Sugar consumption may positively react to income growth at lower income levels (e.g. in Japan) but with higher consumption levels saturation is reached (e.g. in the USA). Butter exhibits either negligible or negative correlation with income in the countries covered here.

Summarizing our discussion on the influence of income on the demand for food we find that, depending on the type of food and the circumstances, food consumption may react positively, negatively or not at all to income growth. However, this rather inconclusive statement can be somewhat refined when we consider Engel's law which says that, even if the demand reaction to income growth is positive, it is generally, at least at somewhat higher incomes, less than proportional.

THE ROLE OF PRICES

In the previous section we have faithfully adhered to the *ceteris paribus* clause when dealing with the influence of income on the demand for food. We shall continue to follow this principle, but now take income and all other factors as

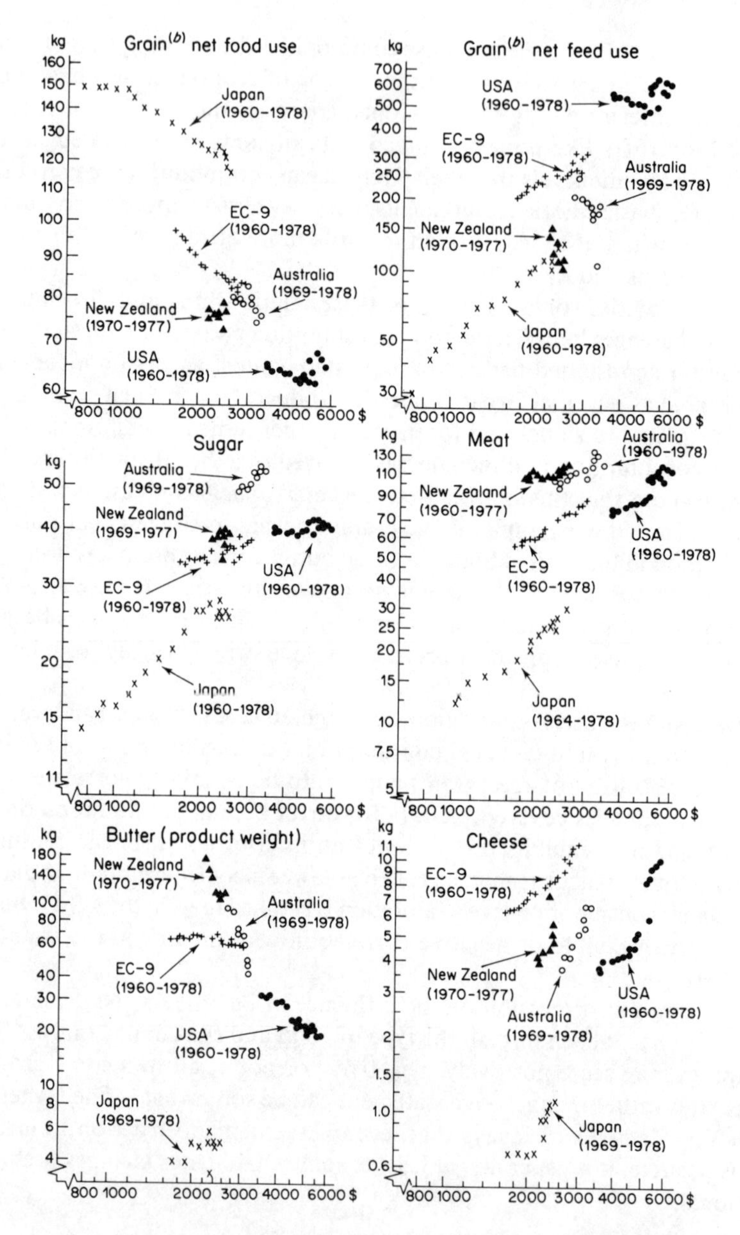

Figure 4.1 Food consumption and GDP per head[a].

a. Food consumption in kg per head and year; GDP in constant prices and exchange rates of 1970, US$ per head; both scales logarithmic.

b. Including rice.

Source: OECD *Food Consumption Statistics,* various issues.
OECD *Main Economic Indicators,* various issues.

given and look into the reactions to price changes. Unfortunately, there is nothing so plausible or universally applicable as Engel's law, as far as price reactions are concerned. Thus, we shall have to put up with a few general hypotheses and rather open conclusions here. Only somewhat later shall we see that Engel's law has some (weak) implications for price reactions as well and that the findings connected with the name of an even earlier empirical researcher, Mr King, are closely related to this.

Let us first consider the reaction of the quantity bought of any commodity to a change of this commodity's price—the 'own-price reaction' in the economist's jargon. Few people would doubt that the sign of this relationship is negative, i.e. if the price rises the quantity bought goes down and vice versa. The explanation appears simple. If a commodity's price goes up the consumer is no longer able, and is possibly also not willing, to buy the original quantity. Can we, then, develop an hypothesis about the magnitude of this reaction? At first glance we could be tempted to think that the quantity reaction is proportional to thc pricc changc, i.c. if thc pricc riscs by 1 pcr ccnt thc quantity bought decreases by 1 per cent. We could argue for this by assuming that the consumer's buying decisions are such that he allocates certain shares of his budget to each individual commodity and then goes out for the shopping. In this case he would divide the money allocated to each commodity by its price and this would give the quantity he can buy. If this were true, the quantity bought of each commodity would, in fact, have to vary proportionally (and inversely) with its price.

However, on reflection we detect that this is not necessarily true, because it would mean that the consumer looks at each commodity separately, without weighing commodities against each other. Take the example of meat consumption. It is rather likely (and has empirically been found to be true in many cases) that, if the price of pork meat goes up, the consumer does not only buy less pork meat but also more—say—chicken meat, because it has now become (relative to pork meat) cheaper. This means that the consumer does not look at each commodity separately but that he may be willing to substitute among commodities. To the extent that this is true, it must have consequences for the magnitude of the reaction to price changes. Taken as such, the existence of substitution among commodities could lead us to expect that the quantity reaction to price changes is more than proportional. If, in reaction to a price rise of pork meat, the consumer buys more chicken meat this could lead us to think that pork consumption has to drop by a higher percentage than the increase in the price of pork meat, because money has to be 'set free' in order to enable the consumer to buy more chicken meat.

On the other hand, it may well be that the consumer is not willing to curtail pork meat consumption so much. He may, rather, prefer to reduce the purchase of other items in order to maintain, in spite of a higher pork meat price, pork meat consumption at close to the original level. In this case the quantity reaction could be considerably less than proportional to the price change. It is, actually, rather likely that an increase of the price of one commodity negatively affects the consumption of many other commodities, even if these other commodities seem to be rather unrelated to the commodity whose price has risen. This is the case because a price increase for any commodity means that the purchasing power of the consumer's income is reduced. By 'purchasing power' we mean, in our context, something like the money income divided by the average price of the goods the consumer buys. If the average price of all goods goes up (because the price of any commodity has risen) the purchasing power of the consumer's income drops. Taken as such this may have the same effect as if, with constant prices for all goods, the consumer's income had been decreased by an equivalent amount. If I spend 20 per cent of my income on housing, and rents go up by 10 per cent, this means that the purchasing power of my income is reduced by 2 per cent. The effect this has on my consumption of all commodities may be similar to a reduction of my income by 2 per cent, with all prices constant.

Considering this, we are now in a position to understand a somewhat more theoretical statement concerning the consumer's reaction to price changes. This reaction can be decomposed into two elements. One we may call the 'income effect'. It is due to the change in the purchasing power of income. Its direction and magnitude can be determined on the basis of our considerations regarding the relationship between income and consumption in the previous section. The second element is called the 'substitution effect'. It results from the changing price ratio between commodities, as discussed above in the example of pork and chicken meat. What we observe when we look into the consumer's behaviour in the market place is, then, always the sum of these two theoretically discernible elements.

This decomposition of the consumer's reaction to price changes is mainly of theoretical interest. However, it is very helpful in explaining one seemingly anomalous but possible reaction to price changes. It is related to the name of the Victorian economist Sir Robert Giffen who was reputed to have observed that an increase in the price of bread led to an increase in the consumption of bread by poor families in nineteenth-century England (Marshall, 1920). This observation would seem to contradict our general expectation that the quantity consumed is inversely related to the price. However, it does not, necessarily. Bread may have accounted for a high share of total expenditure among the families concerned. Hence the increase in the price of bread must have reduced the purchasing power of these families significantly. At the same time, bread may have been an inferior good, such that a decrease in the purchasing power of these families led to an increase in bread consumption. In this case the income effect was positive. The substitution effect may still have had the normal negative sign, indicating that the increase in the bread price relative to other prices would have led the consumers to substitute other food for bread, had they still had the original purchasing power of their income. In other words, the rising bread price made these families so much poorer that they could no longer afford to consume meat but had to resort to even higher bread consumption.

This so-called Giffen case is no longer relevant in today's industrialized countries. However, in some extremely poor developing countries it may still apply to some staple foods which constitute a high proportion of total household expenditure and are, at the same time, inferior.

Summarizing these findings on the relationship between consumption and the own price, we have to conclude that in general anything can happen. If the price goes up consumption can decrease or increase, and it can do so more or less or in the same proportion. However, since the Giffen case is a rare exception, in practically all cases the quantity consumed is negatively related to the price. Moreover, as we shall see below, in most cases we are safe in assuming that, as far as the demand for food is concerned, the quantity reaction is less than proportional to the price change. If this is true, the set of possible reactions is considerably narrowed down.

In this discussion of the own-price reactions we have already touched upon the relationship between consumption of one commodity and the price of a different good when we considered the pork meat and chicken meat case. In this case we found that chicken meat consumption is likely to increase when the price of pork meat goes up. The 'cross price reaction', as the economist calls this relationship, is positive in this case. This is generally true if the two goods considered are substitutes. However, there are other types of commodities which cannot substitute each other but are consumed in combination, a typical textbook example being coffee and sugar (though this example is obviously not valid for all coffee drinkers). For these so-called 'complements', the cross-price effect is negative. If the coffee price increases sugar consumption goes down, because it is now more expensive to drink coffee (with sugar). Thus the cross-price reaction can be positive or negative, depending on whether the pair of commodities concerned have to be considered substitutes or complements.[3] As there is also nothing we can hypothesize about the magnitude of the cross-price effect, no generally true statement can be made about the relationship between one commodity's price and demand for another good.

WHAT IS THE DIFFERENCE BETWEEN ELASTICITY AND FLEXIBILITY?

Let us come back to Engel's law for a moment and let us use it now as the starting point for a somewhat more formal analysis. As we have seen, its original formulation was in terms of the share of food in total expenditure and the way in which it varies with varying family income. If we denote the quantity of food which a family consumes by x, and its price by p, food expenditure is equal to px. Let y be total expenditure (or income, as the distinction between income and expenditure does not concern us here). Then the share of food expenditure in total expenditure is:

$$\frac{px}{y} \text{ or equivalently } p\,\frac{x}{y}$$

Engel's finding that the share of food in total expenditure decreases with increasing income is equivalent, as we have seen above, to the statement that food expenditure increases less than proportionally when incomes rise. In terms of our symbols, this means that x increases less than proportionally with increasing y. Clearly, this implies that the ratio x/y, and hence the share of food in total expenditure, decreases with increasing y. The statement 'less than proportionally' is not very exact, and it is desirable to have a more accurate measure of the way in which the quantity of food consumed, x, changes when income, y, is changing. An obvious candidate for such a measure, given our interest in proportions, is the ratio between the percentage change of the quantity consumed and the percentage change in income from which it results. Let us call this ratio n.

$$n = \frac{\%\text{ change of } x}{\%\text{ change of } y}$$

A measure of this type is a good description of the intensity by which a dependent variable (x) reacts to changes in the casual variable (y). In economics such a measure of responsiveness is called an elasticity, in this case the elasticity of the quantity consumed (or demanded) with respect to income, or in short the 'income elasticity of demand'. Should we observe that, *ceteris paribus*, consumption of food increases by 2 per cent when income increases by 8 per cent we would say that the income elasticity of demand for food is 0.25.[4] Engel's law which, expressed differently, says that the percentage change in x is smaller than the percentage change y in the case of food, obviously implies that the income elasticity of the demand for food is generally below one. In the special case of inferior goods (the demand for which decreases with increasing income) the income elasticity is even below zero (i.e. it is negative).

We have already noted that different types of food respond differently to income changes. Equipped with the elasticity measure, we are now in a position to make more precise statements by referring to empirical estimates of income elasticities for different types of food. Table 4.3 provides a survey of such estimates for selected countries. It tells us, for example, that in France demand for meat would increase by 0.4 per cent if incomes go up by 1 per cent and that the income responsiveness of the demand for meat in France is double that in the USA. We note that cereals tend in fact to be inferior goods (negative

Table 4.3 Income elasticities of demand for selected food items.

Product	USA	Sweden	France	Australia	UK	Italy	Spain	Brazil	Kenya	India	Indonesia
Wheat	-0.3	-0.3	-0.4	-0.1	-0.2	-0.2	-0.3	0.4	0.8	0.5	1.0
Coarse grain	-0.1	-0.3	-0.1	0.0	-0.1	-0.4	-0.1	-0.3	0.4	-0.2	0.4
Sugar	0.1	0.0	0.3	-0.1	0.0	0.4	0.6	0.1	1.0	1.0	1.4
Vegetables	0.1	0.5	0.3	0.2	0.3	0.3	0.5	0.5	0.5	0.7	0.6
Fruit	0.2	0.6	0.5	0.7	0.5	0.6	0.7	0.5	0.5	0.8	0.8
Meat	0.2	0.2	0.4	0.1	0.2	0.7	0.7	0.5	1.0	1.2	1.3
Eggs	-0.1	0.1	0.2	0.0	0.0	0.5	0.6	0.6	1.0	1.0	1.2
Fish	0.3	0.3	0.6	0.3	0.3	0.4	0.7	0.5	0.8	1.5	1.0
Milk	-0.5	-0.2	0.1	-0.1	-0.1	0.3	0.5	0.6	0.8	0.8	2.0
Butter	-0.5	-0.2	0.2	0.0	0.0	0.4	0.5	1.1	0.9	0.6	n.a.
Coffee	0.0	0.3	0.5	0.8	0.8	1.0	1.0	0.1	1.0	0.4	0.2
Farm value	0.04	0.1	0.2	0.1	0.1	0.3	0.4	0.3	0.6	0.6	0.7
GNP per head (1971, $)	5160	4240	3360	2870	2430	1860	1100	460	160	110	80

Source: Ritson (1977).

income elasticities) in most countries, except at lower levels of income, e.g. in Indonesia, and that in general the income responsiveness of higher valued items like meat, fruit and coffee is greater than for other types of food.

Another interesting conclusion we can draw from estimates like those compiled in Table 4.3 is that, in general, the higher the level of income, the lower income elasticities of the demand for food tend to be.[5] This statement can be thought of as an extension of Engel's law (or, in mathematical terms, the first derivative of Engel's law). Engel's law implies that the income elasticity of the demand for food is less than one; its extension is that the elasticity declines with increasing income.

If the elasticity measure proves useful in describing the relationship between income and consumption, why should we not use an equivalent concept for analysing the responsiveness of consumption to price changes. Let us, therefore, define the price elasticity of demand, e, as

$$e = \frac{\text{\% change of } x}{\text{\% change of } p}$$

Hence if in reaction to a price increase of 5 per cent (*ceteris paribus*, of course) the quantity demanded declines by 1 per cent, we would say that the price elasticity of demand is -0.2.

At this point we have prepared the ground for appreciating the findings of another early piece of empirical research into the behaviour of food consumers. It is generally attributed to Gregory King[6] and referred to as 'King's law'. It is based on the observation that the smaller the corn harvest, the higher grain prices become. In particular, the price increase is *more than proportional* to the harvest loss. Obviously, this has implications for the magnitude of the price elasticity of demand for grain. If the harvest is small, prices must go up in order to force consumption down to the level of availability. If the percentage price rise is greater than the corresponding percentage decline in consumption, our definition of e tells us that the implied price elasticity of demand must be less than one in absolute terms (it must be somewhere between 0 and -1). As similar observations have been made for different types of food in many cases, we are quite safe in stating that in general (though not necessarily in each individual case) price elasticities of the demand for food are, in absolute terms (i.e. neglecting the negative sign), below one. In fact, it is quite plausible that food consumption should be rather price-inelastic as food is both a necessity and a saturation good. If food prices go up, we cannot curtail food consumption very much, and if they decline, possible expansion of food demand is also limited.

King's law is even easier to interpret when we relate it to the reciprocal of the price elasticity. Let this be called q.

$$q = \frac{\%\ \text{change of } p}{\%\ \text{change of } x}$$

The interpretation of q is the percentage price change which is necessary in order to bring about a 1 per cent change in consumption. Equivalently, it allows us to calculate the percentage price change which results from a given percentage change in supply. Hence q gives the responsiveness of prices to changing availability of commodities, which is exactly what King's law is concerned with. Economists call q the price flexibility and it is obvious that:

$$\text{price flexibility} = \frac{1}{\text{price elasticity}}$$

If it is true that the price elasticity of demand for food is generally below one (in absolute terms), the price flexibility in food markets must be above one (King's law). Hence relatively small fluctuations in supplies cause relatively large fluctuations in prices. It is, indeed, well known that food markets tend to react rather hectically to changing availability of supplies. Governments of most countries, therefore, feel the need to be heavily engaged in stabilizing food markets.

ARE REAL PRICES REALLY REAL?

In a modern economy, incomes and prices are expressed in money terms (rather than in kind). Natural as this appears, it may cause confusion in analysing the behaviour of consumers (or any other economic agent) if the unit of measurement, money, changes its value during our observation. The value of money, say of 1 dollar, is determined by the amount of goods we can buy with it. The higher prices are in general, the less we can buy with 1 dollar. Inflation, i.e. an increase in the general level of prices, reduces the value of money. Therefore, if money incomes (or 'nominal incomes') grow by, say, 6 per cent in a given year, this does not mean that the purchasing power of consumers has risen by the same percentage, if inflation has taken place. Should the general level of prices have increased during this year by 4 per cent, the purchasing power of the consumers' income, or their 'real income', has grown by only about 2 per cent. Of course, it is only when incomes grow in real terms (as opposed to nominal terms) that consumers are able to buy more goods.

Similarly, a change in a commodity's money (or nominal) price does not tell us very much when we do not know what has happened to all other prices. If the general price level rises by 4 per cent and bread prices increase by only 3 per cent, bread becomes cheaper relative to other goods. The 'real price' of bread in this case decreases by 1 per cent. Again, we would expect that it is only the ratio of a commodity's price to all other prices, i.e. the real price, which matters for the consumer's decision as to what to buy.

So far in our considerations we have not encountered these problems as we have always stuck to the *ceteris paribus* clause. When we considered the influence of income, we assumed that all prices remained constant. Clearly, in this case there is no distinction between nominal and real changes of income. Equivalently, when we discussed the implications of a price change, we kept all other prices constant, so that nominal and real prices were the same.

However, when we want to apply our considerations to real-world statistics we have to take account of the fact that, in reality, many variables change simultaneously. We shall, in our empirical research, probably never be able to include all relevant factors in an analysis of consumer behaviour. However, we want to consider at least the most important factors appropriately. This means that we cannot ignore the existence of inflation whenever we work with time series data. The most direct way of accounting for inflation, in fact, is to convert all money variables, i.e. income and prices, into real variables and to use only these real variables in the analysis. In practice, this 'deflating' is done by dividing all nominal numbers by a price index which reflects the development of the general (or, more accurately, the average) price level. The resulting numbers (e.g. dollar income of 1980 divided by price index for 1980) are then values in real terms which can be used directly in our analysis. Indeed, in an economic analysis, these 'real' numbers make much more sense than the actual numbers we find in statistics.

FOOD DEMAND IN THE LONG RUN

Having discussed the main economic factors influencing the demand for food and concepts by which we can describe their effects quantitatively, we can now try to bring everything together in order to analyse actual developments in food consumption. What we want to do here is to relate changes in food demand over time to the economic factors from which they result. Though the introduction to this chapter has made the point that economic factors certainly do not explain everything in the demand for food, we shall, for the sake of simplicity, argue here as if food consumption were dependent only on incomes and prices. Moreover, we shall confine our discussion to growth rates of food consumption, rather than considering absolute levels.

Imagine we want to predict the development of food consumption (in so far as it is due to economic factors). What information do we need? We need to know (or rather to assume) how income and prices will change in the future and how these changes will affect the demand for food. Let us assume we have reason to believe that, over the period in which we are interested, real incomes per head will grow by an annual rate of 4 per cent. The effect this has on food demand must depend on the income elasticity of demand. Since the income elasticity is defined as percentage change of demand per percentage change of income, we have to multiply the growth rate of real income by the income elasticity in order to determine the income effect on future demand. If the income elasticity is 0.3, food consumption will grow by $0.3 \times 4\% = 1.2\%$ per annum as far as the impact of income growth is concerned.

In exactly the same manner we have to deal with prices. However, for the sake of simplicity let us disregard the cross-price effect and deal with 'own prices' only. We may have the expectation that in the period under consideration the real price of food will decline at a rate of 1 per cent per annum (i.e. that the nominal food price will rise by 1 per cent per annum less than the general price level). If the price elasticity of the demand for food is -0.2, the price effect on consumption growth will be $(-0.2) \times (-1\%) = 0.2\%$ annually. We may then predict that, in the period concerned, food consumption per head will grow by the sum of the income and the price effect, i.e. by $1.2\% + 0.2\% = 1.4\%$ per annum. If we want to predict aggregate consumption growth we have to add to this the expected annual growth rate of population in percentage terms.[7] Easy as it may appear to be to predict food consumption or the demand for any individual food in this way, in practice projections of future consumption growth are not without difficulties. First, there is little information on which expectations concerning future incomes and prices can be based. Second, in spite (or because) of much empirical research into demand elasticities, in each individual case it is difficult to pick the appropriate elasticity values. Moreover, even if we knew the accurate elasticity values, there is no guarantee that they will remain constant in the future. Finally, factors other than incomes and prices may have a significant impact which is not accounted for in our analysis. Yet, having said this, it is reassuring to note that many studies based essentially on this approach (though with many refinements) have been rather successful in predicting future food consumption.

In order to practice this approach, and at the same time to gain some insights into the long-term development of the demand for food, let us consider the way in which the growth of food demand changes during the process of economic development. Table 4.4 provides an illustration of the way in which the factors considered here may affect the demand for food in three different stages of development. Of course, this is a highly schematic description of what are in fact rather complex interrelations, but it may still gave some idea of the major forces at work.

In early stages of development, income growth is relatively low, but the income elasticity of demand for food is comparatively high. The growth of food supply may just about keep pace with demand growth such that real food prices neither rise nor fall. A slight growth in population adds to the income effect, and aggregate growth of food demand may be slightly below 2 per cent p.a. During the phase of industrialization, income growth may be considerable. Though the income elasticity is already somewhat lower, the income effect on food demand can be quite high. As demand growth may be well ahead of supply expansion, real food prices may tend to increase, which results in a slightly negative price effect. On the other hand, population growth can be considerable at this stage, such that the aggregate growth rate of food can be as high as 4 per cent p.a. or above. In mature economies, finally, growth rates of both income and population tend again to be low, income and price elasticities are low as well,[8] and even slightly declining real prices for food (which reflect a tendency of productivity growth in modern agriculture to outstrip the growth of food demand) do not add significantly to demand expansion, which may be as low as 1 per cent p.a. or less.

Table 4.4 Schematic illustration of the growth of food demand at different stages of economic development.

	Stage of development					
	Pre-industrial		Industrialization		Maturity	
Growth rate of real income per head p.a.	1%		4%		2%	
Income elasticity	0.7		0.5		0.2	
Income effect		0.7%		2.0%		0.4%
Growth rate of real food prices p.a.	0%		1%		-1%	
Price elasticity	-0.6		-0.4		-0.4	
Price effect		0.0%		-0.4%		0.1%
Growth rate of population p.a.		1.0%		2.5%		0.5%
Growth rate of food demand p.a.		1.7%		4.1%		1.0%

In this schematic illustration, 'food' stands for the aggregate farm value of the raw products contained in the various food items the consumer buys at the retail level. This is not because of any conceptual characteristics of the approach applied here which works for any type of food at any level (and, indeed, for any good in general). It is only the magnitude of the elasticities used in Table 4.4 which is supposed to represent the aggregate value at farm level of the raw products embodied in total food consumption.[9] Obviously, this is one particular way of aggregating over all the various food items consumers buy. A different—and possibly more obvious—way of aggregation is to add up the value of food consumption at the retail level.[10] What would the difference be between these two ways of defining 'food'?

In the second section of this chapter we have already mentioned the major point of difference between the development of food demand at the farm level and that at retail level. It is the relatively rapid growth of demand for 'processing and distribution' which results in an increasing gap between the retail and the farm-level value. Table 4.5 illustrates this development using German data. The share of farm-level value in food value at the retail level varies considerably among different types of food, reflecting different amounts of goods and services which are combined with the raw product before it reaches the consumer. Generally, the percentage marketing margin (i.e. the difference between the value at retail level and the value at farm level) is higher for vegetable products than for animal products. However, for most product categories covered, the percentage margin exhibits a clearly increasing trend (or, equivalently, the share of farm-level value a decreasing trend).

Table 4.5 Percentage share of farm sales in consumer expenditure on food of domestic origin, West Germany.

Product	1964-66	1972-74	1979-81
Bread and cereal products	18.5	14.0	12.1
Potatoes	70.2	58.5	47.0
Sugar	39.2	41.9	41.4
Vegetables	31.2	30.8	37.8
Fruit	51.6	49.7	35.3
All vegetable products	32.5	29.5	23.0
Meat and meat products	53.2	50.8	45.2
Milk and dairy products	62.9	56.6	58.1
Eggs	86.7	85.3	80.1
All animal products	58.0	54.7	50.7
All animal and vegetable products	50.8	48.4	43.6

Source: *Agrarbericht der Bundesregierung*, various issues.

Farmers' lobbies occasionally use statistics of this type when arguing that farm prices are too low and that this is due to obscure forces working in the processing and distribution sector, which deprive farmers of their legitimate share of the money consumers spend on food. However, it is easy to understand that this development is a typical symptom of the process of economic growth. In a way we can interpret it as a particular corollary of Engel's law. Food as such, i.e. the physical raw material contained in food, is necessarily a good which is subject to saturation. However, 'processing and distribution' is a 'good' of which increasing amounts can easily be consumed. When incomes rise, consumers prefer convenience potato products to raw potatoes, lean meat to fat (which means that processing meat becomes more expensive), deep frozen products to canned food, nicely packed and presented food to bulk goods and so on. This means that the income elasticity of the demand for 'processing and distribution' is greater than that for the raw product content of food. Expressed in terms similar to Engel's law, this implies that the share of farm value in consumers' expenditure on food tends to decrease with increasing income.[11]

Table 4.6 presents an example of an attempt at estimating the different magnitudes of the income elasticities of demand for raw products at the farm level on the one hand and for processing and distribution on the other hand. Though estimates of this type are difficult and the results have to be interpreted with caution, this example may serve to illustrate that income elasticities of demand for processing and distribution can be quite high.

Table 4.6 Estimates of income elasticities of demand for agricultural raw products at farm level and for processing and distribution, Denmark, 1970-1976.

	Income elasticity of demand for:	
	Raw product at farm level	Processing & distribution
Wheat	0.03	
Rice	-0.01	
Coarse grains	-0.06	
Processing and distribution of bread and cereals		0.53
Bovine and ovine meat	0.42	
Pork meat	0.21	
Poultry meat	0.73	
Processing and distribution of meat		2.15
Butter fat	0.06	
Milk fat (other than butter)	0.37	
Milk protein	-0.12	
Eggs	0.12	
Processing and distribution of dairy products and eggs		0.98
Fruit	0.60	
Vegetables	0.16	
Processing and distribution of fruit and vegetables		0.45

Source: Haen, Murty and Tangermann (1982).

CONSUMPTION ANALYSIS WITHOUT ECONOMIC THEORY?

In our simple description of the way in which economic factors like income and prices influence the demand for food we have largely done without theory. We have started from observable behaviour of the 'typical' or 'average' consumer (e.g. observations described by Engel's and King's laws), tried to give plausible explanations of this behaviour (e.g. by referring to the saturation phenomenon) and discussed systematic ways of describing this behaviour (by expressing it in terms of elasticities). We have not, however, presented a formal analysis of the consumer's decision procedure, by which his consumption behaviour may be determined.

The economics literature is full of formal modes of consumer behaviour.[12] The most basic model starts from the notion that, with given prices and income, the consumer is able to buy various combinations of quantities of all goods. The second element of this theoretical model is a well defined preference order which enables the consumer to rank all relevant bundles of goods. Confronting the set of all attainable combinations of quantities of goods with this preference order, the model consumer can then pick that bundle of goods which ranks highest within this constrained set (which provides the highest 'utility' he can attain). This is the model of the homo economicus referred to above. This theoretical person has full information of all possible courses of action and rationally decides in favour of that which maximizes his objectives.

There are many refinements of this basic theory. For example, the act of consumption can conceptually be split up into two sets of activities, one being the buying of goods in the market place and the other being the use of these goods in the household. The consumer can then be interpreted as an economic agent who combines market goods, time and 'human capital' in order to produce utility-yielding non market goods. In this view it is no longer the goods themselves which are relevant for the consumer's decision but certain 'characteristics' of these goods. Developments of this approach have, therefore, successfully been applied to the analysis of the implications of quality changes in consumer goods.

This type of economic theory may appear rather esoteric, and in many ways it is. However, to economists it has proven rather useful in two respects. First, it has served to structure our ideas of the way in which various factors may interact when consumers make decisions. This aspect is important mainly in a didactic sense. Second, it has greatly helped to develop methods by means of which we have been able to pursue systematic empirical research into the economics of consumer behaviour.[13]

This aspect is of great relevance for the type of issues we have discussed in this chapter. Though we have not explicitly dealt with the formal economic theory of consumer behaviour in this chapter, much of this theory was implicitly behind many of the explanations we have given, and some of the empirical data which have been presented have been produced by methods which were based on applications of this theory.

It is difficult fully to appreciate the usefulness of economic theory for empirical research without going into an area which we have not touched upon in this short chapter—the area of econometrics. Econometrics is the art of skilfully combining economic theory and statistical analysis in order to pursue fruitful empirical research into economic phenomena. In a nutshell, it culminates in formulating equations which describe the way in which given economic variables (say quantities demanded) depend on explanatory variables (say income and prices), and in estimating the numerical values of the parameters in these equations on the basis of observed data. Some of the data presented in this chapter have resulted from such econometric work (e.g. Tables 4.3 and 4.6). The role of economic theory is decisive in econometrics as the formulation of the estimation equations is (or at least should be) based on sound theory.

Having said this, it should be clear that our discussion in this chapter of the economic factors influencing the demand for food has necessarily remained rather superficial. We have tasted some results of economic theory, but not really digested any theory itself. However, on occasion tasting whets the appetite, and this chapter would fully serve its purpose if it would induce some readers to dig deeper into the economic theory of consumer behaviour.

NOTES

1. Moreover, as in other cases, inconsistencies in published statistics may spoil the picture. This could, for example, be the reason for the surprising break in the time series for US cheese consumption.
2. Except feed grain in Australia, which may be due to the fact that the statistics include feed use for exported animal products as well, such that changes in exports may distort the numbers on feed use.
3. In passing, we have to note that the same pair of goods can be substitutes for one consumer and complements for another. Take the case of schnaps and beer. In some German regions they are considered complements as it is customary always to drink a schnaps before having a beer ('in order to warm the stomach'). In other regions they are deemed substitutes as people would have either schnaps or beer and would consider it dangerous to have them in combination (which it actually is, according to this author).
4. Note that elasticities have no dimension. This is very convenient as it facilitates comparison between goods, countries and periods. No matter what the unit of measurement of x (kilogram, pound, litre, etc.) and y (£, $, DM; per month, per year; etc.) is, by way of expressing everything in terms of percentage we eliminate dimensions and gain comparability.
5. Note that in Table 4.3 countries are arranged in descending order of their level of income as measured by Gross National Product (GNP) per head in 1971.
6. Though it obviously originated from a study by Charles Davenant published in 1699. See Stigler (1954, pp.95-113).
7. By simply adding up percentage growth rates in this way we make a slight numerical mistake as we should, also, include the product of the growth rates. For example, if consumption per head grows by 2 per cent and population by 1 per cent aggregate consumption grows by 3.02 per cent rather than by 3 per cent. However, as long as the percentages we deal with are relatively small, the magnitude of this mistake is negligible.
8. Note that in Table 4.4 income and price elasticities have been changed in parallel from stage to stage. This is to reflect one basic result of the theory of demand which requires that for any good the sum of the income elasticity, the own-price elasticity and all cross-price elasticities equals zero. Hence if the income elasticity declines with rising incomes (see our extension of Engel's law above), the own-price elasticity has to decline as well, if the cross-price elasticities do not change very much.
9. See the equivalently defined income elasticities in the last but one line of Table 4.3.
10. In both cases we would have to use constant prices of a chosen base year as weights for aggregation.

11. This statement applies to the 'quantity' of processing and distribution consumed. It therefore describes the quantity element. Prices of goods and services included in the margin tend to rise faster than farm level prices of agricultural raw materials. This is mainly due to different rates of productivity growth. Hence the price element, too, adds to the expansion of the margin. Numbers like those presented in Table 4.5 clearly represent the sum of these two elements.
12. For an up-to-date account of this branch of economic theory see, for example, Deaton and Muellbauer (1980).
13. For a presentation of the way in which economic theory is applied to the empirical analysis of consumer behaviour see, for example, Phlips (1974).

REFERENCES

Burk, M.C. (1962) 'Ramifications of the relationship between income and food', *Journal of Farm Economics,* **XLIV**, 1 February.

Deaton, A., and Muellbauer, J. (1980) *Economics and Consumer Behaviour.* Cambridge.

Eurostat (1983) *Yearbook of Agricultural Statistics 1978-1981.* Brussels.

Haen, H. de, Murty, K.N., and Tangermann, S. (1982) Kunftiger Nahrungsmittelverbrauch der Europaischen Gemeinschaft—Ergebnisse eines simultanen Nachfragesystems. Munster-Hiltrup.

Marshall, A. (1920) *Principles of Economics*, 8th edn. Macmillan, London.

Phlips, L. (1974) *Applied Consumption Analysis.* North Holland, Amsterdam, Oxford, New York.

Ritson, C. (1977) *Agricultural Economics; Principle and Policy.* Collins, London.

Stigler, G.J. (1954) 'The early history of empirical studies of consumer behaviour', *Journal of Political Economy,* **62**.

The Food Consumer
Edited by C. Ritson, L. Gofton and J. McKenzie

CHAPTER 5

Psychological Factors Influencing Food Choice

PAUL ROZIN, MARCIA LEVIN PELCHAT AND APRIL E. FALLON

INTRODUCTION

Food plays a central role in our lives as a source of nutrition and emotional gratification and as a vehicle for expression of social relations and values. Yet, we know little about the nature and origin of people's preferences, likes and attitudes. This lack of knowledge is due in large part to the fact that psychology, arguably the science most appropriate to explore these issues, has been fascinated with the issue of what determines *how much* people (and animals) eat rather than *what* they eat. Although the study of this field is in its infancy, psychologists have taken a number of different approaches to the study of food selection. The traditional approach has been to examine what causes a food to be accepted or rejected, and to ask questions like: What is the role of biology in the formation of food habits? Are there innate taste and odour preferences? Can some individual differences in food habits be traced to early learning experiences? What is the role of culture in shaping individual food habits?

The answers to these questions form a varied and interesting literature and we shall consider possible answers to some of them in this chapter. As most researchers would agree, it is difficult to separate the cultural or social dimensions of food choice from psychological or, indeed, biological factors. Although these are dealt with elsewhere in this volume (see Murcott, McKenzie, Gofton) we here include reference to those factors which seem germane to the general project of producing a psychology of food choice.

A less traditional approach to the study of food habits focuses on the psychological categories into which food acceptances and rejections fall. For example, instead of asking, 'How did Fred come to reject liver?' or 'Did he become ill after eating it when he was a little child?', we ask, 'Does Fred reject liver because it looks and smells bad, because he believes that it is dangerously high in cholesterol or because "organ meats" disgust him?' In this chapter, we will focus on the psychological categorization of foods, and using a format which we (Rozin and Fallon, 1980) have developed, we will briefly consider how biology, culture and individual experience may affect whether a food will be accepted or rejected.

A TAXONOMY OF PREFERENCES AND AVOIDANCES

What gets into the mouth is influenced less by the physical properties of objects (e.g. the nutrients they contain) than by the psychological properties the individual attributes to them. In this section, we will analyze the motivational (psychological) bases for acceptance or rejection of foods. To do that we must first distinguish three very different terms which are often used with each other; these are 'use', 'preference' and 'liking'. The *use* of a food item refers to what and how much of it is actually consumed by a person or group. *Preference* assumes a situation of choice, and refers to which of two or more foods is chosen. One might 'prefer' steak to bread but eat ('use') more bread because of price, or availability. *Liking*, refers to an *affective* response to foods, and is one determinant of preference. A dieter might 'prefer' (choose) lettuce to cake but like cake better. Now these concepts are clearly related. All other things being equal, we eat (use) what we prefer, and we prefer what we like. However, while availability, price and convenience are important determinants of use, they do not determine preference or liking; while perceived health value of a food can be a potent determinant of preference and use (Krondl and Lau, 1982), it may have little to do with liking. For example, Woodward (1945), in his study of an attempt to increase acceptance of soya products in cafeterias, suggests that an emphasis on the nutritive value of new foods implies that they do not taste good.

Table 5.1 Psychological categories of acceptance and rejection.

	Rejections				Acceptances			
Dimensions	Distaste	Danger	Inappropriate	Disgust	Good taste	Beneficial	Appropriate	Transvalued
Sensory affective	—			—	+			+
Anticipated consequences		—				+		
Ideational		?	—	—		?	+	+
Examples	Beer, Chili, Spinach	Allergy foods, Carcinogens	Grass, Sand	Feces, Insects	Saccharine	Medicines	Ritual foods	Leavings of heroes or deities[a]

[a]As in consumption of items offered in ritual sacrifice.
Source: Fallon and Rozin (1983).

The distinctiveness of preference, use and liking argues for the importance of looking systematically at what motivates people to consume or reject foods. By early adulthood, every individual in every culture has adopted a culturally based set of beliefs and attitudes about objects in the world, with respect to their edibility. We have explored this psychological categorization of substances in American culture and have concluded that there are at least three basic types of reasons for acceptance or rejection of potential food objects (see recommended reading at end of chapter). Each of these reasons (Table 5.1) in one form motivates acceptance, and in the opposite form motivates rejection. This simplified system emphasizes the principal feature motivating acceptance or rejection (although many foods are rejected or accepted for more than one reason). Each of these reasons is considered below.

FOOD EVALUATION DIMENSIONS

Sensory Affective

Some substances are rejected or accepted primarily because of their sensory effects in the mouth; that is, because of their taste, texture, odor and sometimes appearance. In our terms, accepted items fall into the psychological category of 'good tasting' and rejected items into the category of 'distaste'. Good tastes and distastes, by definition, respectively elicit appropriate positive and negative affect (like or dislike). Sensory-affective reaction may be innately based (such as acceptance of sweet tastes and rejection of bitter ones) or acquired. Substances which fall into the sensory-affective category for any individual are almost always acceptable food in his culture. Individual differences in this category (e.g. liking or disliking spinach) probably account for most of the variation in food preferences *within* a culture.

Anticipated Consequences

Some substances are accepted or rejected as food primarily because of anticipated consequences of ingestion. These consequences could be rapid effects (such as nausea or cramps, or a pleasant feeling of satiation) or more delayed effects involving belief and attitudes about the health value of substances (such as vitamins or low-fat foods on the positive side, or potential carcinogens on the negative side). Anticipated consequences need not be psychological: they may also be social, such as expected changes in social status as a consequence of eating a food. All acceptances based on anticipated consequences wc place in the 'beneficial' category, and all rejections in the 'dangerous' category. Note that for this category, liking for the taste is not an important dimension; if a liking (disliking) is present, it would be considered an additional cause of acceptance (rejection).

Ideational

Some substances are rejected or accepted primarily because of our idea or knowledge of what they are or where they come from. While ideational factors predominate in many food rejections, they are less common in acceptance. We distinguish between two distinct subcategories of ideational rejection: 'inappropriate' and 'disgust'.

Items classified as 'inappropriate' are considered inedible within the culture and are refused simply on this basis. Grass and paper are examples. These items are not presumed to taste bad. They are usually emotionally neutral and are not in any way offensive.

In contrast, the category of 'disgust' carries a strong negative emotional response. In our analysis, disgusting items consistently elicit nausea, and are thought to have unpleasant tastes, although in most instances they have never been tasted. There is also a negative affective response to their odor and appearance, and to skin contact or the presence of the offending substance in the body. Items viewed as disgusting are 'contaminants' or 'pollutants'; that is, the possibility of their presence in a food, even in the tiniest amounts, makes the food unacceptable (although this is sometimes true of dangerous foods as well). A few substances, such as feces, seem to be universally disgusting, in our terms. In the USA and the United Kingdom, the category includes insects and the meat and/or viscera of a variety of animals (e.g. dogs).

DETERMINANTS OF FOOD CHOICE

For analytic convenience, determinants of food choice can be categorized as biological, cultural or individual (psychological). If one were interested in determining as much as possible about an adult's food preferences and attitudes, and could only ask him/her one question, the question would surely be: what is your culture or ethnic group? In the process of enculturation, exposure to particular foods is controlled and food values and attitudes are conveyed. These issues are considered in depth by Murcott. There are also many differences within cultures in food preferences and attitudes. Since we do not know what is responsible for these differences, we could not readily address a very informative second question to our hypothetical person. We will try in this chapter to suggest some causes of some of the unexplained variation by examining some of the influences on individuals' preferences for foods.

BIOLOGICAL FACTORS IN FOOD SELECTION

Any discussion of biological influences on human food habits must take account of the fact that humans are omnivorous; their basic biological make-up, with a general purpose dentition and digestive system, allows them to exploit a wide variety of nutrient sources. The adaptive value of omnivory rests on this ability to exploit a wide variety of food sources rather than depending upon one potentially unreliable source. It is not possible to specify in advance what sensory (visual, tactual, olfactory, taste, etc.) properties will distinguish foods from inedible items. For example, some plants are edible while other, related, species contain substances toxic to many animal species. The individual omnivore must be able to learn what is edible.

This is not to deny that there are biological determinants of food choice. Evidence for biological (genetically determined) guidelines would include an innate (present at birth) preference for sweet tastes and avoidance of bitter tastes (Steiner, 1977; Cowart, 1981); an interest in new foods, coupled with a fear of them (Rozin, 1976a); and some special abilities to learn about the delayed postingestional consequences of foods (Garcia *et al.*, 1974; Rozin, 1976a; Booth, 1982).

Curt Richter (1943) demonstrated that rats have the ability to choose safe, nutritionally sound diets. Clara Davis (1928) made a similar claim for 'wisdom of the body' in humans. In a well-known study, Davis offered three children, for a period of months to years immediately after weaning, an array of about a dozen foods to choose from at each meal. She reported that the children showed normal growth, even though what they selected from the range made available to them was completely under their own control. These are important findings, but they do not establish a case for a biological wisdom of the body in humans. All of the foods offered to the children were of good nutritional value, so that random choice among them would probably have led to normal growth. No refined products (e.g. sugar) or flavorings were added to the foods. Most critically, the preferred foods of the children were milk and fruit, the two sweetest choices available. We do not know what they would have done if sweet-tasting desserts had appeared on the menu.

Taste Preferences and Digestive/Metabolic Influences

There seems to be a universal, biologically based attraction to sweet, and avoidance of bitter, substances. These responses are innate: newborn infants make positive facial expressions and accept sugar water placed on their tongues, and make negative facial expressions and reject bitter tasting water placed on their tongues. Studies of bottle-fed infants in the USA showed that sugar enhances intake and it is added to infant foods in many cultures, presumably for this reason. Interestingly, bad tastes do not seem to motivate rejection by human or rat neonates if they are encountered during *suckling* (Desor *et al.*, 1975). This 'nipple protection' has obvious adaptive value for infants in traditional cultures who have no alternative food source to mother's milk. However, the infant's unresponsiveness to bad tastes delivered by a nipple can also have maladaptive consequences. This is most poignantly illustrated by a case where salt was inadvertently substituted for sugar in infant formula in a hospital; tragically, despite a salty taste, the liquid was nevertheless ingested in large enough quantity to result in death. (Finberg *et al*, 1963) The discrepancy between responses to tastes placed on the tongue and tastes encountered through suckling appears to wane as the age of weaning approaches, because bitter or irritant substances rubbed on the nipple are effective in promoting weaning.

Individual differences in food choice can also have a biological basis. Studies have revealed genetically based differences in sensitivity to some bitter compounds in humans. Fisher *et al* (1961) report weak relations between bitter sensitivity and preferences for bitter foods. Changes in taste and smell sensitivity with age studied by Schiffman (1979) and Cowart (1981) may account for small changes in preference as the years go by.

Digestive and metabolic differences among individuals and between ethnic groups may account for differences among cultural practices (Simoons, 1982; Katz, 1982). A case in point is lactose intolerance. Most humans today (and presumably all humans prior to the origin of dairying) cannot digest lactose (milk sugar) as adults (Simoons, 1978). Moderate amounts of milk produce gas, bloating and diarrhoea. In addition, most of the sugar in milk, and some of the calcium, is not absorbed. Adults who come from cattle-herding, milk-drinking traditions, such as Northern Europeans, can digest lactose. There is strong evidence that lactose tolerance in adulthood is an inherited trait and that, across cultures, the use and acceptance of milk is related to the degree of lactose tolerance in the population.

The current physiological or metabolic state of a person is also a potent determinant of food choice. The clearest case is the state of satiety. It is common knowledge that people choose different foods when hungry or sated. When hungry they learn to consume more calorically rich foods than when sated (Booth, 1982). In addition, the pleasantness of sweet tastes varies with state of repletion (Cabanac, 1971).

The effect of other metabolic factors on preference is better documented in animals than in humans. For example, rats show a prompt increase in intake of sodium during periods of sodium need (Richter, 1956; Denton, 1982). This response is not learned but appears immediately the first time a sodium-deficient rat encounters a source of sodium. Sodium appetite has been reported in humans as well. Most notable is a case reported by Wilkins and Richter (1940) of a young child with chronic sodium deficiency which resulted from adrenal cortical insufficiency. From the age of 12 months, this child showed a strong preference for salty foods and refused to eat other foods unless salt was added to them. 'Salt' was one of his first words. He died, tragically, shortly after admission to the hospital (for other symptoms) because his need for sodium was not recognized, and the hospital diet did not provide him with adequate amounts of the mineral.

Other 'specific hungers' have been demonstrated in the rat. These, in contrast to sodium appetite, appear to consist of a combination of weak learned preferences for nutrient-rich diets and stronger learned aversions to deficient diets (see Rozin, 1976a for a review). There are numerous anecdotal and case history reports of cravings or 'specific hungers' in humans but there is little hard evidence on this topic. Davis (1928) reports ingestion of massive quantities of cod liver oil by a child with rickets (vitamin D deficiency). Pica (clay, starch or earth eating) may be associated with iron deficiency anemia or other mineral deficiencies (Cooper, 1957) and food cravings during pregnancy (Trethowan and Dickens, 1972) may result from changes in internal (hormonal) state. However, many cravings associated with deficiency states do not appear to be particularly adaptive. Olynyk and Sharpe (1982) report on a woman suffering from iron deficiency who experienced an intense craving for paper. She reported eating, on average, a box of paper tissues and a cigarette package per day. She also admitted to eating an entire paperback book in a day. Not only did paper ingestion fail to provide her with the iron that she needed, but it resulted in serious mercury poisoning (mercury is used in paper processing).

Variety and Familiarity

Interest in and simultaneous fear of new foods is presumably another part of our omnivorous heritage. This appears as a desire for variety and at the same time a preference for familiar foods. Highly monotonous diets, where the same cycle of meals is repeated every few days, lead to a decrease in the palatability of foods involved. According to Rolls *et al.* (1982) there is also a more immediate effect, called sensory-specific satiety, where the palatability of a given food drops within one meal, with respect to other, similar items that have not been consumed in that meal. Presumably as a consequence of this, when people are offered a variety of items (e.g. different sandwiches or different desserts) they eat more in one meal than they do of the single most favored item when it is served alone.

A feature of human food selection that is as salient as the avoidance of monotony is the way that most people narrowly circumscribe the range of acceptable foods. Many people are reluctant to try new foods. Even in a society as ethnically varied as that of the United States, and one without its own well-defined cuisine, many foreign cuisines (e.g. the food of India) have not been accepted, even as occasional curiosities. Meanwhile, the fast-food chains of the country, providing a highly predictable, invariable and narrow range of foods, prosper, at least in part because in this case Americans, but probably people in general, seem to like the fact that they know exactly what they are getting. The desire for comfort and familiarity in eating is not surprising: eating involves putting things into the body, a very personal act, and one surely does not want the emotional importance and pleasure of eating to be marred by fears and uncertainties.

We have given examples of avoidance both of new foods and of monotonous diets. People seem to prefer a moderate level of variety. New foods, once tried, tend to become liked better. However, this assumes that unfamiliar foods are sampled in the first place and virtually nothing is known about why individuals differ in their willingness to sample new foods or about how to overcome unwillingness to sample new foods.

There are wide individual differences in receptivity to new foods. A striking example of this is seen in the United States in development by many young children (particularly in the 2-4 year age range) of behavior involving rejection of almost all foods (Bakwin and Bakwin, 1972). Although the few foods consumed by such children may form an adequate diet, the behavior is very disturbing to parents. We do not know the role of constitutional factors and early experience in this common syndrome. There are, however, some promising leads. Animal studies suggest that the amount of dietary variety experienced early in life may influence later willingness to sample novel foods (Capretta *et al.*, 1975). Preliminary work by Pelchat and Pliner (1986) which is consistent with these results suggests that young children may resemble their parents in willingness to try new foods. This may come about because parents who commonly try new foods may offer their children a wider variety of foods.

LEARNING ABOUT FOODS

There seems to be a specific adaptation which aids omnivores in learning to avoid dangerous potential foods. When ingestion of a food is followed by illness (usually nausea), humans (and rats) develop a strong aversion to the food (Garcia *et al.*, 1974; Rozin and Kalat, 1971; Garb and Stunkard, 1974; Bernstein, 1978). This learning can occur with delays of hours between ingestion and illness experience. Flavors are more likely to be associated with illness than with peripheral pain, while illness is also more likely to be associated with flavors than with the appearance of food, or with places (e.g. restaurants). These taste aversions have an 'irrational' character. People acquire aversions to foods where ingestion was accompanied by illness, even when they 'know' that the food was not responsible for the illness (e.g. in cases where illness was caused by the flu, or motion sickness) (Logue, Ophir and Strauss, 1981; Pelchat and Rozin, 1982).

Most learning about food, of course, comes about not through trial and error or through the realization of innate dispositions, but through socialization. People everywhere are members of society and, in consequence, both heir to and participants in the constitution of the culture, or way of life which is distinctive to the group of which they are members.

Now, culture influences food choice and use in a wide variety of ways (Garine, 1971). Specific systems of food production have been implemented in different parts of the world to exploit specific resources or eco-systems, and this has led to the establishment of cuisines based on food resources with particular limits.

Food preparation, processing and flavoring at the same time show an amazing ingenuity and diversity as the human need for variety and change struggles to transcend these limits by invention. Food, for people everywhere, is clearly not just about nutrition but also serves social purposes, and is used in celebration, in worship, in communication and even in competition with others (Douglas and Nicod, 1974).

All societies have socially defined rules for food use, and specific concepts of what is, or is not, to count as food. While these tend to foster conservatism of cuisine, and to inhibit innovation, this is not to say that food habits do not change. People also want variety, and seek to avoid monotony in their diets. The identification of areas in which changes can be made is a problem, of course, for those who want to improve the nutritional status of diets, or to introduce new products. We have already argued that predictability, safety and familiarity are the foundation on which the markets for fast food are built.

In an important sense, we can see culture interacting with biology to constrain or predispose food choice. The nexus for this interaction is, of course, the individual, who develops a unique set of preferences and attitudes. Some analysts have argued that culture reflects biological factors to a high degree—indeed, that cultural practices with regard to food have a biologically adaptive value (Katz, 1982; Harris, 1974; Simoons, 1982) The best example is that alluded to already; lactose intolerance in adults has led to the development in dairying cultures of food products in which the lactose is 'digested' externally in the form of cheese or yoghurt, or else the non-lactose elements of milk products are separated off by processing, as in ghee in India.

Attempts to identify biologically based behavioral aspects of selection are less obvious; the widespread use of sweets, and the panoply of food products developed in this area, can be pointed to as deriving from what seems to be a basic urge for this type of taste. But clearly there are innumerable instances of tastes which are acquired in spite of a 'natural' aversion (spices such as chili pepper, coffee, alcohol, tobacco, to name only a few). Here culture has overcome biological predisposition.

Both biological and cultural factors are important elements in food choice, interacting with psychological factors. It is to the psychology of the individual that we now turn.

INDIVIDUAL PSYCHOLOGICAL FACTORS IN FOOD SELECTION

Here we will discuss how individuals' experience with foods affect their attitudes toward foods, and the mechanisms by which social and cultural influences are translated into individual behavior. We will do this first by considering whether there is a special role for early experience in establishing food preferences and then by considering the processes involved in developing likings for foods.

Early Experience

It is a commonly held assumption that early experience has a strong effect on lifelong food preferences. There is no question that early experience can influence food habits. However, it is not clear that early experience is a more potent force in shaping lifelong preferences than is later experience.

It is difficult to separate the effects of early and later experience on human food habits since (especially in traditional cultures) the child is exposed to the same food repertoire for his/her entire life. Thus, most of the information about early experience comes from animal research. There is some evidence, in non-mammals, of a special role for early experience in the formation of food preferences. For example, snapping turtles (Burghardt and Hess, 1966) and chickens (Capretta, 1969) preferred their first food to an equally familiar one experienced later.

Young mammals also acquire preferences for specific foods experienced early in life (Galef, 1977). These preferences, however, do not appear to be particularly strong or long lasting. So, although these studies demonstrate an effect of early experience on food preferences, they do not demonstrate that there are special mechanisms for the effects of early experience on lifelong food preferences. In most of these studies, familiar, first foods were compared to novel foods. Thus, early experience effects in mammals are probably based on a mechanism that is common to young and adult mammals: a tendency to prefer familiar foods to novel foods (Galef, 1977). In humans, the strong preference for milk (Rozin, 1976b) might be seen as evidence for an early experience effect since almost all humans, early in life, have a diet which consists of warm milk. However, given that warm, not cold, milk is our earliest food, it is surprising that warm milk is liked no more than cold milk by preschool children (Herbert-Jackson *et al.*, 1977) and is strongly disliked by many adults (Rozin, 1976b).

The fact that individuals can acquire strong food preferences as adolescents or as young adults provides further evidence against a special early experience effect in human food preferences. Good examples of this are the strong attachments that are formed to alcoholic beverages, coffee and chili pepper even if these foods are not introduced into the diet until adolescence.

In retrospect, we probably should not be surprised that there is so little evidence for a special effect of early experience. The key to an omnivore's success is its ability to exploit a wide variety of foods and therefore to avoid falling prey to fluctuations in a single food source. It would therefore be particularly maladaptive for rats or humans—omnivores *par excellence*—to become wedded for life to their earliest foods (Rozin, 1976b). Furthermore, the first food of all mammals is milk, and it would certainly be unwise for an animal to become rigidly attached to this substance in infancy since (with the exception of some humans) it never encounters it again once it is weaned.

More promising than the idea that early experience produces particularly strong attachments to particular foods is a relatively neglected line of evidence which suggests that early experience affects a more general aspect of food selection: willingness to accept novel foods. It has been demonstrated that animals which experienced early dietary restriction show greater resistance to the acceptance of novel foods later in life than do animals whose early diets were varied (Kuo, 1967; Capretta *et al.*, 1975). Evidence in humans is less conclusive. An interview study with the mothers of preschool children has revealed a relationship between the amount of variety in children's diets and children's willingness to accept novel foods (Pelchat and Pliner, 1986). However, while it is plausible that children who receive varied diets become more accepting of novel foods, it is also possible that children who are more interested in novel foods get offered more varied diets.

Acquisition of Likes and Dislikes

Of particular interest to us here is the mystery of how some objects are felt to have good or bad tastes, while others are simply accepted or rejected on the basis of anticipated consequences or ideational factors. The acquisition of liking or disliking is of special interest in the study of food selection for two reasons:

(1) Although we can understand that someone avoids a food because he has been told it will make him sick, we do not understand what causes him to come to dislike or like its taste.
(2) From the point of view of public health, it would be highly desirable for people to like what is good for them, and dislike what is bad. Acceptance of a food because it tastes good is far more stable than acceptance of a food because it is nutritious. This is probably because the food is then eaten for itself, rather than for some extrinsic reason.

We will consider three contrasting pairs of categories (from Table 5.1). In each case, the first member of the pair involves an affective (like or dislike taste) response, and the second does not. The contrasts will be:

(1) distasteful versus dangerous foods;
(2) good tasting versus beneficial items; and
(3) disgusting versus inappropriate items.

Distaste versus Danger

There is an innate aversion to bitter foods, but there are many items that are initially acceptable and later come to be rejected, some because we come to experience them as tasting bad, others because we learn that they are dangerous (see discussion of learned taste aversions, above). What determines whether an object will acquire one or both of these properties?

Nausea and vomiting are particularly potent in causing acquired distastes when they follow ingestion of a particular food (Pelchat and Rozin, 1982). When other negative events such as hives, respiratory distress, headache and cramps follow eating, the induced avoidance is motivated by danger rather than by distaste. This contrast can be illustrated by typical cases of two individuals who avoid strawberries. One has an allergy to strawberries and gets skin symptoms or life-threatening respiratory distress after eating strawberries. Such a person will avoid strawberries as dangerous but will probably continue to like their taste. If his allergy could be treated, he would be delighted to consume strawberries. The other person originally liked strawberries but got sick and vomited after eating them. This person will typically dislike the taste of strawberries and avoid them though realizing that they are, in fact, not 'dangerous', and were perhaps not even the cause of the sickness.

These results suggest that the presence or absence of nausea is one critical factor that determines whether a food will be distasteful. There are limits to the explanatory power of the distaste-nausea linkage. It seems very unlikely that most acquired distastes have a history of association with nausea. Not even half of the people surveyed (Garb and Stunkard, 1974; Logue *et al.*, 1981) could remember even one instance of a food-nausea experience, and these same people had a great many distastes. So we must presume that the many people who dislike lima beans, fish, broccoli, etc., have reasons as yet undiscovered, and there are unfortunately no sound theories as to what other paths to distaste might be.

Although adults can perceive the same food as distasteful and dangerous, they can clearly separate these features. In contrast, young children confound these features. Many American children in the 4-7-year range apparently believe that if a food is harmful, it also tastes bad (Fallon *et al.*, 1984).

Good Taste versus Beneficial

The acquisition of likes for food

Although there is much more research and information about acquired likes than about dislikes, no single factor (as potent as nausea for dislikes) has been identified. The most consistent and prominent empirical relation in the study of acquired likes in all domains is that exposure tends to increase liking. This 'mere exposure' effect (Zajonc, 1968) has been demonstrated for foods by Pliner (1982), who reported increased liking with increased exposure to fruit beverages. 'Mere exposure', however, does not always facilitate liking (see, for example, Peryam, 1963) and the reverse can occur with overexposure to the same food (Stang, 1975). The literature (Harrison, 1977; Berlyne, 1970) suggests that mere exposure should be most effective where moderate levels of exposure to novel and 'complex' foods are experienced.

A fundamental question that remains is whether *mere* exposure itself is ever sufficient, or whether exposure allows some other process to occur. In either case, other factors do contribute to the acquisition of likings. Let us list and evaluate some possibilities, and consider what underlines the distinction between good tastes and beneficial items.

Given the strong linkage between a particular physiological effect (nausea) and distaste, we might expect to find a physiological effect which is especially potent in producing likes. Booth (1982) found that hungry subjects come to prefer (and increase their liking for) the flavor of a high-calorie food to that of a low-calorie food. He suggests that flavors (foods) whose ingestion is followed by satiation often become liked. However, these changes in liking depend upon the state of satiety of the individual. This type of liking may be somewhat different from what we could call 'stable' likes; that is, there is an important sense in which shrimp are preferred to celery whether one is hungry or not.

A second possible mechanism of acquired liking is the association of a neutral or disliked food (flavor) A with an already liked food B. The simplest form of association would be offering A and B together, or following ingestion of A by that of B. In a recent laboratory study, one flavor was served to young adults in a sweet (palatable) beverage, while a second flavor was served equally frequently in an unsweetened, less palatable form. There was an enhanced liking for the first flavor, even when both flavors were served in the unsweetened form (Zellner *et al.*, 1983). We cannot yet define the real-life circumstances under which this type of conditioning might take place. The effectiveness of these pairings would certainly be constrained by culture (e.g. it is unlikely that a new meat could be made more acceptable by covering it with whipped cream).

In omnivorous animals it is much harder to establish preferences than aversions (Rozin and Kalat, 1971; Zahorik, 1979) and this has been particularly true for innately unpalatable (e.g. bitter or irritant-spicy) foods. In contrast to animals, humans develop strong preferences for many foods, including spices and coffee, despite their innate unpalatability (Rozin, 1982). This difference suggests that there are some important factors that are operating only in humans. Sociocultural influences are natural candidates.

We believe that social factors operate at two levels. First, society (custom, the behavior of elders, the foods made available to the child) essentially forces exposure, and as we have pointed out, exposure fosters liking. Second, the perception that a food is valued by respected others (e.g. parents) may itself be a mechanism for establishment of liking.

Children show increased preferences for foods preferred by peers (Birch, 1980a), heroes and elders (Duncker, 1938; Marinho, 1942). Birch *et al.* (1980) have shown that preference for a snack is enhanced if it is given to the child in a positive social context, or still better, if it is used by the teacher as a reward. It is the teacher's indication that she values the food (she is using it as a reward) that seems to be critical. If the same food is given to other children, at the same frequency, in a non-social context (e.g. if it is left in their lockers), there is no enhancement of preference.

The importance of the perception of social value in acquired liking is emphasized by the converse phenomenon. How can an object of intrinsic value (which we call a liked object) lose some of this value? Work by Lepper (1980) and Birch *et al.* (1982) suggests that the perception by the child that others do not value the food *per se*, and that they must be bribed into eating it (e.g. rewarded, told how healthy it is) destroys the intrinsic value of it (e.g. reduces the liking for the food). For example, when preschool children were rewarded for eating a particular food they increased ingestion while the reward was in effect, but decreased their intake below baseline after the reward was terminated.

The notion that reward decreases or perhaps blocks liking is consistent with the fact that people are less likely to come to like oral medicines than foods, since medicine ingestion is clearly motivated by anticipated beneficial consequences (Pliner *et al.*, 1985). These findings suggest that if an individual believes that he is eating something for its own sake his liking may be increased, but that if eating is governed by some extrinsic factor liking will be decreased.

The reversal of innate aversions: Chili pepper

Any satisfactory theory of the acquisition of acquired likes must be able to account for the development of liking for innately unpalatable substances (e.g. alcohol, tobacco, coffee and irritant spices); they are among the more popular foods around the world. We have been investigating how people come to like the initially distasteful burn of chili pepper. First of all, there is no doubt that people who consume chili pepper like it for the very same reason that others dislike it: the 'mouth burn'. Capsaicin, the active pungent agent in chili pepper, is a physiologically active chemical (Rozin, 1978): it stimulates the gastrointestinal system, causing salivation, increased gastric secretion and gut motility. The salivation enhances the flavor of the frequently bland and mealy diets that are eaten in association with chili pepper, and this, along with the pairing of the pepper with the satiation produced by the accompanying food may contribute to the acquisition of liking. In addition to these possibilities, we are inclined to emphasize the potency of social factors. Study of the acquisition of liking for chili pepper in a Mexican village suggests that chili eating is not explicitly rewarded, but is acquired as an 'imitation' of the behavior of respected adults, older children and peers (Rozin and Schiller, 1980). The amount in the diet is gradually increased for children from 3 to 4 years old, and by 5-7 years of age most children have developed a liking for it.

Ironically, initial dislike may be what is critical. We present here two possible explanations of the acquired liking for chili (and perhaps other innately unpalatable substances) that depend on the initial negativity.

The mouth pain produced by chili may become pleasant as people realize that it is not really harmful. This puts the pleasure of eating chili pepper in the category of thrill seeking (or 'benign masochism'), in the same sense that the initial terror of a roller coaster ride or parachute jumping is replaced by pleasure. People may come to enjoy the fact that their bodies are signalling danger but their minds know there really is none (Rozin and Schiller, 1980).

Alternatively, the many painful mouth experiences produced by chili may cause the brain to attempt to modulate the pain by secreting endogenous opiates—morphine-like substances. There is evidence that, like morphine, these brain opiates do reduce pain. At high levels, they might produce pleasure. Hundreds or thousands of experiences of chili-based mouth pain may cause larger and larger brain opiate responses, resulting in a net pleasure response after many trials (Rozin, 1982).

Family influences on food preferences

Since much evidence in the personality literature indicates the importance of the parents in the formation of the child's personality, habits and preferences, and we have suggested that behavior of elders influences preferences, we might expect to find family resemblances in food preferences. It is surprising that when the effects of culture are removed, studies have found either no, or a very small, relationship between children's preferences for specific substances and those of their parents (see Birch, 1980b, for a review of all but the most recent studies). Although family resemblance could be caused by genetic factors or experience, we are inclined to attribute most of the reported resemblances to experience. In all the studies reporting significant parent-child resemblances, the children were of college age (over 17 years of exposure to parents) (Pliner, 1984; Rozin *et al.*, 1984). Almost all of the negative findings are with young children (2-7 years of exposure) (e.g. Birch, 1980b). Furthermore, mother-father resemblances in preferences are equal to or higher than parent-child correlations. The mother-father correlations tend to increase with number of years married, indicating an effect of mutual influence and exposure to the same items (Price and Vandenberg, 1981).

Multiple mechanisms and motivation for acceptance

We are very far from an understanding of the development of liking for foods. The issue is complicated by two facts. First, many substances are consumed for multiple reasons (Russell, 1974). For example, in the case of coffee, motivations for drinking include liking the taste, anticipated positive effects of caffeine, avoiding caffeine withdrawal, associated social interaction, 'warming up', satiation and so on. Second, our experience in trying to understand the liking for chili pepper convinces us that there are multiple routes to liking. We can find convincing arguments against every single mechanism we have been able to suggest. This indicates not that these mechanisms play no role in acquisition, but that there may be more than one way in which liking is acquired for particular items.

Disgust versus Inappropriate

All previous categories discussed probably exist in both animals and humans. These two ideationally based categories, however, require the mediation of culture, which classifies inappropriate items (e.g. grass, rubber bands) as non-foods. This is learned by the child along with other information about the substance. 'Disgusting items' are also rejected primarily because of the idea of what they are. However, the acquisition of information about disgusting items is linked with a strong negative affective response. This affective response is critical in distinguishing 'disgusting' from 'inappropriate' items; disgusting items are offensive, while inappropriate items are essentially neutral. The offensiveness extends to the idea of the disgusting substance in the body and to its sensory properties: taste, smell and appearance (Rozin and Fallon, 1980; Fallon and Rozin, 1983). Even the thought of consuming a disgusting substance can lead to nausea. Objects of disgust are treated as contaminants; that is, a physical trace of a disgusting item can render an otherwise liked food undesirable (trace contamination). For example, most Americans would refuse a liked beverage if they believed there was even a microscopic trace of urine in it. Although danger may be postulated as a reason for this response, even when any possible physical trace is absent, contact with disgusting items may motivate rejection (pure associational contamination). For example, many adults will not eat a favorite soup if it has been stirred with a *brand-new* fly swatter (Rozin *et al.*, 1984).

The nature of the objects falling into inappropriate and disgust categories also differs. Disgusting objects are almost always of animal origin (Angyal, 1941; Rozin and Fallon, 1980). Most tabooed foods (which we believe would usually have disgust properties) are animal or animal products (Angyal, 1941; Tambiah, 1969). Inappropriate items, by contrast, are usually vegetable or inorganic in nature.

Angyal (1941) suggests that disgust is a fear of oral incorporation of an offensive substance. This is consistent with the facial disgust response, which seems designed to keep offensive substances from the nose and mouth. In general, the more intimate the contact with a disgust substance (and the more real the threat of incorporation), the greater the disgust response (Fallon and Rozin, 1983).

Although we do not as yet have a satisfactory explanation of the origin of disgust, we will put forth some hypotheses on the basic nature of disgusting objects (Rozin and Fallon, 1981). One view is that, at its core, disgust is an inborn aversion to spoiled and decaying matter. This could be related to the fact that decaying food may be more likely to contain pathogenic organisms. However, there is little evidence to support this idea. In fact, young children are attracted to decay odors (Petó, 1936). This attraction disappears in the 2-7-year age range.

A second view focuses on feces as the primary disgust substance but assumes that this disgust is acquired. The universality of disgust for feces in adult humans argues for this primacy (Angyal, 1941). Given the initial attraction, or at least curious neutrality, to feces, the emotionally laden toilet training experience seems the natural source for the origin of disgust. The conversion of an attraction into a strong aversion (disgust) is a paradigmatic instance of what Freud described as a reaction formation (Senn and Solnit, 1968). The generalization of disgust to other objects may be related to the perceptual or conceptual similarity of these objects to feces.

A third view emphasizes filth and disorder. Association with filth (e.g. garbage, feces) would be a means for generalization of the category. Douglas (1966), taking a more symbolic approach, identifies 'filth' as disorder, anomaly, or 'matter out of place'.

A fourth position views animalness as central. Angyal (1941) notes that as the animalness is removed from the offensive object (e.g. cooking, chopping, skinning) the object becomes less disgusting. Others have commented on the particular features of animals that lead to disgust, such as resemblance to humans (as in primates) and relation to humans (as in pets).

Finally, disgust can be related to the social system, where it can be seen as an emotionally laden aspect of human social relations. Many objects of disgust are human products or involve some human mediation. Food is a major means of social expression as can be seen in the use of food in India (Appadurai, 1981). Incorporation of food produced, touched or handled by specific others can involve 'incorporation' of these others. Here the value of food would depend on the nature of the relation to that other (Appadurai, 1981; Meigs, 1978). Our study on family resemblances in food attitudes (Rozin *et al.*, 1984) provides evidence for the importance of lasting social influence in the development of disgust. In fact, the most substantial intracultural parent-child resemblance was disgust sensitivity (the degree of concern for cleanliness of foods and the affective response to contact with and ingestion of substances labelled by members of the cultural group as disgusting).

While each of these views may contain part of the truth, no single one can explain the variety of disgusts that individuals manifest. Except for the innate spoilage view, all entail an acquisition or enculturation process.

We do know that by the age of 4 children in the USA show a strong rejection for adult disgust substances (Fallon *et al.*, 1984). However, the reasons they give for rejection are danger and distaste. Until the age of about 8, they do not show one of the critical signs of disgust; the contamination response. That is, they will willingly consume juice after a disgusting item has been removed from it.

These findings do not negate any of the views on the origin of disgust; we have not resolved this question but only presented possibilities. We do not understand the process through which some items become disgusting, although we presume that the offensiveness is conveyed to children via verbal and non-verbal emotional expressions of parents and respected others.

CONCLUSION

In this chapter we have explored how humans come to select their foods by presenting a psychological categorization of potential foods, and we have discussed the role of biology, culture and individual experience in determining whether a substance is accepted or rejected as a foodstuff.

We have asked many questions and have given few definitive answers. This is because multiple factors influence choice and attitudes, and also because there is a dearth of research in the area of the psychology of food choice. We look forward to major advances in this area within the next few decades.

REFERENCES AND FURTHER READING

Angyal, A. (1941) 'Disgust and related aversions', *J. Abnorm. Soc. Psychol.*, **36**, 393-412.

Appadurai, A. (1981) 'Gastropolitics in Hindu South Asia', *Am. Ethnologist*, **8**, 494-511.

Bakwin, H., and Bakwin, R. M. (1972) *Behavior Disorders in Children.* W. B. Saunders, Philadelphia.

Bavly, S. (1966) 'Changes in food habits in Israel', *J. Am. Diet. Assoc.*, **48**, 488-495.

Berlyne, D. E. (1970) 'Novelty, complexity and hedonic value', *Perception and Psychophysics*, **8**, 279-286.

Bernstein, I. L. (1978) 'Learned taste aversions in children receiving chemotherapy', *Science*, **200**, 1302-1303.

Birch, L. L. (1980a) 'Effects of peer models' food choices and eating behaviours on preschooler's food preferences', *Child Dev.*, **51**, 489-496.

Birch, L. L. (1980b) 'The relationship between children's food preferences and those of their parents', *J. Nutr. Educ.*, **12**, 14-18.

Birch, L. L., Birch, D., Marlin, D. W., and Kramer, L. (1982) 'Effects of instrumental consumption on children's food preference', *Appetite*, **3**, 125-134.

Birch, L. L., Zimmerman, S. I., and Hind, H. (1980) 'The influence of social-affective context on the formation of children's food preferences', *Child Dev.*, **51**, 856-861.

Booth, D. A. (1982) 'Normal control of omnivore intake by taste and smell'. In *The Determination of Behaviour by Chemical Stimuli*, ed. J. Steiner and J. Ganchrow. Information Retrieval, London, 233-243.

Booth, D. A., Mather, P., and Fuller, J. (1982) 'Starch content of ordinary foods associatively conditions human appetite and satiation, indexed by intake and eating pleasantness of starch-paired flavours', *Appetite*, **3**, 163-184.

Burghardt, G. M., and Hess, E. H. (1966) 'Food imprinting in the snapping turtle, *Chelydra serpintina*', *Science*, **151**, 108-109.

Cabanac, M. (1971) 'Physiological role of pleasure', *Science*, **173**, 1103-1107.

Capretta, P. J. (1969) 'Establishment of food preferences in the chicken (*Gallus gallus*)', *Anim. Behav.*, **17**, 229-231.

Capretta, P. J., Petersik, J. T., and Stewart, D. J. (1975) 'Acceptance of novel flavours is increased after early experience of diverse tastes', *Nature*, **254**, 689-691.

Cooper, M. (1957) Pica. Thomas, Springfield, Ill.

Cowart, B. J. (1981) 'Development of taste perception in humans. Sensitivity and preference throughout the lifespan', *Psychol. Bull.,* **90**, 43-73.
Crosby, A. W. (1972) *The Columbian Exchange. Biological and Cultural Consequences of 1492*. Greenwood Press, Westport, Conn.
Davis, C. (1928) 'Self-selection of diets by newly-weaned infants', *Am. J. Dis. Child.,* **36**, 651-679.
Denton, D. (1982) *The Hunger for Salt*, Springer-Verlag, New York.
Desor, J., Maller, O., and Andrews, K. (1975) 'Ingestive responses of human newborns to salty, sour and bitter stimuli', *J. Comp. Physiol. Psychol.,* **89**, 966-970.
Desor, J., Maller, O., and Turner, R. (1973) 'Taste in acceptance of sugars by human infants', *J. Comp. Physiol. Psychol.,* **84**, 496-501.
Douglas, M. (1966) *Purity and Danger*. Praeger, New York.
Douglas, M., and Nicod, M. (1974) 'Taking the biscuit: the structure of British meals', *New Society*, 19 December, 744-747.
Duncker, K. (1938) 'Experimental modification of children's food preferences through social suggestion', *J. Abnorm. Soc. Psychol.,* **33**, 489-507.
Fallon, A. E. and Rozin, P. (1983) 'The psychological bases of food rejections by humans', *Ecology of Food and Nutrition,* **13**, 15-26.
Fallon, A. E., Rozin, P., and Pliner, P. (1984) 'The child's conception of food. The development of food rejections, with special reference to disgust and contamination sensitivity', *Child Development,* **5**, 566-575.
Finberg, L., Kiley, J., and Lutrell, C. N. (1963) 'Mass accidental salt poisoning in infancy', *J. Am. Med. Assoc.,* **184**, 121-124.
Fischer, R., Griffin, F., England, S., and Garn, S. M. (1961) 'Taste thresholds and food dislikes', *Nature,* **191**, 1328.
Galef, B. G. (1977) 'Mechanisms for the social transmission of acquired food preferences from adult to weanling rats'. In *Learning Mechanisms in Food Selection,* ed. L. M. Barker, M. R. Best and M. Domjan. Baylor University Press, Waco, Texas, 123-148.
Garb, J. L., and Stunkard, A. (1974) 'Taste aversions in man', *Am. J. Psychiatry,* **131**, 1204-1207.
Garcia, J., Hankins, W.G., and Rusiniak, K. W. (1974) 'Behavioural regulation of the milieu interne in man and rat', *Science,* **185**, 824-831.
Garine, I. de (1971) 'The socio-cultural aspects of nutrition', *Ecology of Food and Nutrition,* **1**, 143-163.
Goode, J., Theophano, J., and Curtis, K. (1981) 'Group-shared food patterns as a unit of analysis'. In *Nutrition and Behaviour*. ed. S. A. Miller. Franklin Institute Press, Philadelphia, 19-30.
Grivetti, L. E., and Paquette, M. B. (1978) 'Nontraditional food choices among first generation Chinese in California', *J. Nutr. Educ.,* **10**, 109-112.
Harris, M. (1974) *Cows, Pigs, Wars and Witches: The Riddles of Culture*. Random House, New York.
Harrison, A. A. (1977) 'Mere exposure'. *In Advances in Experimental Social Psychology*, vol. 10, ed. L. Berkowitz. Academic Press, New York.
Herbert-Jackson, E., Cross, H. Z., and Riley, T. R. (1977) 'Milk types and temperature—what will young children drink?', *J. Nutr. Educ.,* **9**, 76-79.
Jerome, N. (1977) 'Taste experience and the development of a dietery preference for sweet in humans: ethnic and cultural variations in early taste experience'. In *Taste and Development. The Genesis of Sweet Preference*, ed. J. M. Weiffenbach. (DHEW Publication No. NIH 77-1068). U.S. Government Printing Office, Washington, DC, 235-245.

Katz, S. (1982) 'Food behaviour and biocultural evolution'. In *Psychobiology of Human Food Selection*, ed. L. M. Barker. AVI, Westport, Conn., 171-188.

Krondl, M., and Lau, D. (1982) 'Social determinants in human food selection'. In *Psychobiology of Human Food Selection*, ed. L. M. Barker. AVI, Westport, Conn.

Kuo, Z. (1967) *The Dynamics of Development: An Epigenetic View*, Plenum, New York.

Lepper, M. R. (1980) 'Intrinsic and extrinsic motivation in children: detrimental effects of superfluous social controls'. In *Minn. Symp. Child Psychol., Vol. 14*, ed. W. A. Collins. Lawrence Erlbaum, Hillsdale, NJ, 155-214.

Logue, A. W., Ophir, I., and Strauss, K. E. (1981) 'The acquisition of taste aversions in humans', *Behav. Res. Ther.*, **19**, 319-333.

Marinho, H. (1942) 'Social influence in the formation of enduring preferences', *J. Abnorm. Soc. Psychol.*, **37**, 448-468.

Mead, M. (ed.) (1943) 'The problem of changing food habits', *National Research Council Bulletin*, **108**, October, 1-177 (whole volume).

Mehren, G. L. (1966) 'Geography, geopolitics and world nutrition'. In *Food and Civilization*, ed. S. M. Farber, N. L. Wilson and R. H. L. Wilson. C.C. Thomas, Springfield, Ill., 121-131.

Meigs, A. S. (1978) 'A Papuan perspective on pollution', *Man*, **13**, 304-318.

Mintz, S. (1979) 'Time, sugar, and sweetness', *Marxist Perspectives*, **2**, 56-73.

Olynyk, F., and Sharpe, D. H. (1982) 'Mercury poisoning in paper pica', *New Engl. J. Med.*, **306**, 1056-1057.

Pelchat, M. L., and Pliner, P. (1986) 'Antecedents and correlates of feeding problems in young children', *J. Nutrition Education*, **18**, 23-29

Pelchat, M. L., and Rozin, P. (1982) 'The special role of nausea in the acquisition of food dislikes by humans', *Appetite*, **3**, 341-351.

Peryam, D. R. (1963) 'The acceptance of novel foods', *Food Technology*, **17**, 33-39.

Pető, E. (1936) 'Contribution to the development of smell feeling', *Br. J. Med. Psychol.*, **15**, 314-320.

Pliner, P. (1982) 'The effects of mere exposure on liking for edible substances', *Appetite*, **3**, 283-290.

Pliner, P. (1983) 'Family resemblance in food preferences', *J. Nutr. Educ.*, **15**, 137-140

Pliner, P., Rozin, P., Cooper, M., and Woody, G. (1985) 'Role of medicinal context and specific post-ingestional effects in the acquisition of liking for the tastes of foods'. *Appetite*, **6**, 243-252.

Price, R. A., and Vandenberg, S. G. (1980) 'Spouse similarity in American and Swedish couples', *Behav. Genet.*, **10**, 59-71.

Richter, C. P. (1943) 'Total self regulatory functions in animals and human beings', *Harvey Lect.*, **38**, 63-103.

Richter, C. P. (1956) 'Salt appetite of mammals: its dependence on instinct and metabolism'. In *L'Instinct dans le comportement des animaux et de l'homme*, ed. F. S. Polignac. Masson, Paris, 577-629.

Rolls, B. J., Rolls, E. T., and Rowe, E. A. (1982) 'The influence of variety on human food selection and intake'. In *The Psychobiology of Human Food Selection*, ed. L. M. Barker, AVI, Westport, Conn., 101-122.

Rozin, E. (1982) 'The structure of cuisine'. In *The Psychobiology of Human Food Selection*, ed. L. M. Barker. AVI, Westport, Conn., 189-203.

Rozin, E. (1983) *Ethnic Cuisine: The Flavor Principle Cookbook*, Stephen Greene, Brattleboro, Vt.

Rozin, E., and Rozin, P. (1981) 'Culinary themes and variations', *Natural History*, **90**(2), 6-14.

Rozin, P. (1976a) 'The selection of food by rats, humans and other animals'. In *Advances in the Study of Behaviour, Vol. 6,* eds. J. Rosenblatt, R. A. Hinde, C. Beer, and E. Shaw. Academic Press, New York, 21-76.

Rozin, P. (1976b) 'Psychobiological and cultural determinants of food choice', in *Dalhem Workshop on Appetite and Food Intake*, ed. T. Silverstone. Dalhem Konferenzen, Berlin, 285-312.

Rozin, P. (1978) 'The use of characteristic flavourings in human culinary practice'. In *Flavor: Its Chemical, Behavioral and Commercial Aspects*, ed. C. M. Apt. Westview, Boulder, Colo., 101-127.

Rozin, P. (1979) 'Preference and affect in food selection'. In *Preference Behaviour and Chemoreception*, ed. J. H. A. Kroeze. Information Retrieval, London, 289-302.

Rozin, P. (1982) 'Human food selection: the interaction of biology, culture and individual experience'. In *Psychobiology of Human Food Selection,* ed. L. M. Barker. AVI, Westport, Conn., 225-254.

Rozin, P., and Fallon, A. E. (1980) 'The psychological categorization of foods and non-foods: a preliminary taxonomy of food rejections', *Appetite,* **1**, 193-201.

Rozin, P., and Fallon, A. E. (1981) 'The acquisition of likes and dislikes for foods'. In *Criteria of Food Acceptance: How Man Chooses What He Eats*, eds. J. Solms and R. L. Hall. Forster, Zurich, 35-48.

Rozin, P., Fallon, A. E., and Mandell, R. (1984) 'Family resemblances in attitudes to foods', *Developmental Psychology,* **20**, 309-314.

Rozin, P., and Kalat, J. W. (1971) 'Specific hungers and poison avoidance as adaptive specializations of learning', *Psychol. Rev.,* **78**, 459-486.

Rozin, P., and Schiller, D. (1980) 'The nature and acquisition of a preference for chili pepper by humans', *Motivation and Emotion,* **4**, 77-101.

Russell, M. A. H. (1974) 'The smoking habit and its classification', *Practitioner,* **212**, 791-800.

Schiffman, S. (1979) 'Changes in taste and smell with age: psychophysical aspects'. In *Sensory Systems and Communication in the Elderly (Aging, Vol. 10)*, eds. J. M. Ordy and K. Brizzee. Raven Press, New York, 227-246.

Senn, M. J. E., and Solnit, A. J. (1968) *Problems in Child Behaviour and Development*. Lea and Febiger, Philadelphia.

Siegel, P. S., and Pilgrim, F. J. (1958) 'The effect of monotony on the acceptance of food', *Am. J. Psychol.,* **71**, 756-759.

Simoons, F. J. (1978) 'The geographic hypothesis and lactose malabsorption: a weighing of the evidence', *Dig. Dis.,* **23**, 963-980.

Simoons, F. J. (1982) 'Geography and genetics as factors in the psychobiology of human food selection'. In *The psychobiology of human food selection*, ed. L. M. Barker. AVI, Westport, Conn., 205-224.

Stang, D. J. (1975) 'When familiarity breeds contempt, absence makes the heart grow fonder: effects of exposure and delay on taste pleasantness ratings', *Bull. Psychonomic Soc.,* **6**, 273-275.

Steiner, J. E. (1977) 'Facial expressions of the neonate infant indicating the hedonics of food-related chemical stimuli'. In *Taste and Development: The Genesis of Sweet Preference*, ed. J. M. Weiffenbach. (DHEW Publication No. NIH 77-1068.) U.S. Government Printing Office, Washington, DC, 173-188.

Tambiah, S. J. (1969) 'Animals are good to think and good to prohibit', *Ethnology,* **8**, 423-459.

Trethowan, W. H., and Dickens, G. (1972) 'Cravings, aversions and pica of pregnancy'. In *Modern Perspectives in Psych-Obstetrics*, ed. J. S. Howells. Bruner/Mazel, New York, 251-268.

Wilkins, L., and Richter, C. P. (1940) 'A great craving for salt by a child with cortico-adrenal insufficiency', *J. Am. Med. Assoc.,* **114**, 866-868.
Woodward, P. (1945) 'The relative effectiveness of various combinations of appeal in presenting a new food: soya', *Am. J. Psychol.,* **58**, 301-323.
Zahorik, D. (1979) 'Learned changes in preferences for chemical stimuli: Asymmetrical effects of positive and negative consequences, and species differences in learning'. In *Preference Behaviour and Chemoreception*, ed. J. H. A. Kroeze. Information Retrieval, London, 233-246.
Zajonc, R. B. (1968) 'Attitudinal effects of mere exposure', *J. Pers. Soc. Psychol.,* **9** (Part 2), 1-27.
Zellner, D. A., Rozin, P., Aron, M., and Kulish, C. (1983) 'Conditioned enhancement of human's liking for flavour by pairing with sweetness', *Learning and Motivation,* **14**, 338-350.

The Food Consumer
Edited by C. Ritson, L. Gofton and J. McKenzie

CHAPTER 6

You Are What You Eat: Anthropological Factors Influencing Food Choice

ANNE MURCOTT

INTRODUCTION

At some time or other each of us has been astonished by foreign food habits. The British disdain European willingness to eat horsemeat, dub the French 'frogs' and cannot understand their relish for snails. Travellers' tales (and anthropologists') have long reported what strike us as even stranger tastes. Lévi-Strauss (1955) records how the Nambikwara Indians of South America show affection by delousing one another's hair—and then eat each louse that is caught. According to Bascom (1951) the Yoruba of West Africa eat guinea-pigs and Gelfland (1971) explains how the Shona considered locust to be quite a delicacy. Gathered before dawn, while the unsuspecting insects were still asleep, they were then baked with a little water and salt. Even today in Southern African townships, all sorts of specially roasted mice and caterpillars are on sale.

Such strange food habits are rarely a response to shortage or emergency. The inhabitants of Leningrad under siege were reduced to eating cats and mice, and the survivors of an air crash trapped in the Yukon seem to have resorted to cannibalism, but unusual eating habits are not necessarily in the best interests of survival. The Japanese prize the raw flesh of a poisonous puffer fish, potentially so lethal that only specially licensed chefs are allowed to prepare it: despite such precautions, a handful of deaths are still registered each year.

Why do some peoples choose to eat what others avoid? Are there similar patterns of food choice that crop up in otherwise different societies? What do different habits of food preparation and cooking look like? Are the selection of utensils and the choice of menus significant? Are table manners enforced? With whom can one suitably share a meal? In short, what can be learned from the study of food practices, both familiar and unfamiliar? And what have social anthropologists to offer to the search for influences on different eating habits?

We shall see that the answers to these questions centre on a question of *group identity*—a sense of belonging to some particular human group. Of course, one reason for eating is the satisfaction of a physiological need. Food also serves to provide important psychological requirements and, of course, sheer enjoyment. Yet such factors do not wholly account for systematic differences in food choice or for the way people limit selection of items from the much wider, potentially nourishing, variety available. As we shall see, the anthropologist as well as the economist, the psychologist and the sociologist, has something to offer. Eating and drinking are also *cultural* affairs, and it is the study of culture which is the particular province of the anthropologist. Anthropology as a discipline has largely developed around the study of the repository of knowledge, skills, activities, beliefs—in short, the way of life which is passed on from generation to generation within a human group. While cultures, so defined, may differ, culture itself is a universal human characteristic, and a property of human groups alone.

This chapter starts by looking at cooking. This is one of the most distinctive cultural features of human eating and its study shows the way anthropologists link interpretations of actual food practices with analyses of cultural identity. The next three sections each take a case of food avoidance and trace the links that account for them. Common to all three is the significance of examining a group of peoples' way of thinking about and ordering their social and natural environment. Human beings do not simply display a 'reflex' response to their surroundings as animals do, but they make sense of their environment and seek to explain the working of the world they inhabit—and one of the key ways they do this is through food. The following section looks at the association between children's choices of sweets and contrasting adult views of 'proper' eating; and before we come to the concluding section we consider further notions of proper eating, this time in relation to domestic obligations and arrangements. But we begin with 'the raw and the cooked'.

RAW OR COOKED

> . . . I had found a perfect definition of human nature, as distinguished from the animal . . . . My definition of Man is 'a cooking animal'.
>
> James Boswell,
> *The Journal of a Tour to the Hebrides with Samuel Johnson*

Cooking is a universal human practice. However diverse food choice may be worldwide, all known people transform much of what they eat in, on or against the fire. It may be directly by roasting; or indirectly, wrapped in leaves or kitchen foil for baking; or in a saucepan or clay cooking pot of water to be boiled. On the whole, animals consume their food in the state they find or capture it, much of the time eating it on the spot. Thus, when ethologists discovered apes who had learned to rinse their food in the river they were intrigued precisely because such seemingly 'cultured' behaviour is so unusual in animals. Human beings, however, not only wash, peel or wipe their food, not only preserve it by drying, smoking or freezing, they routinely also heat it. What then could be the significance of cooking as a peculiarity of the human species?

The French anthropologist Claude Lévi-Strauss (1966) has taken a close interest in this familiar fact of life. For him, the matter is linked to the further question: what is peculiar about the human species? His answer focuses attention on a paradox. As he reminds us, we all know that human beings have a dual character. As a species we are animal, more or less like other animal species; we have parallel bodily functions and similar biological requirements for survival. At the same time, however, we are not animal but human. We have language, intelligence and culture blended in a way that is unique on this planet. We are creatures of 'Nature' and 'Culture' simultaneously. Moreover, knowledge of this universal feature of human beings is itself universal and forms part of the 'cultural awareness' of all societies. Yet it is also a sources of latent tension in human affairs which, Lévi-Strauss suggests, perennially demands resolution.

It may take a half-conscious and obviously metaphorical form in the case of identifying unusual eating behaviour; we talk of people who 'wolf' their food, have manners 'like a pig' or an appetite 'no better than a sparrow's'. Such responses to food are considered worrying or bad mannered to the point of being uncivilized, effectively un-human. We know that, like animals, we all have to eat, but at the same time we tell ourselves we can do better than animals, and at least know how to eat 'properly'. For Lévi-Strauss, however, the matter can be quite unconscious but is ever-present, none the less, in all aspects of culture, even in the custom of cooking.

We are all familiar with the way that styles of cooking can have connotations, not only gastronomically, but also socially, such that the preparation of food is suited to the occasion or to the sort of person who is to consume it. Roasting meat is 'for' Sundays, banquets and weddings; stewing for the routine meals of a working day. (The association of roasting, in particular, with important events is, Lévi-Strauss claims, to be found in widely differing parts of the world.) By contrast Edwardian cookery books recommend steamed and coddled foods as appropriately bland for the sick; likewise, Hong Kong Chinese residents in London perpetuate the laborious practice of retaining the essential properties of *bo*, i.e. 'repairing' foods that restore strength, by double steaming in a jar with a double lid, itself set in boiling water (Wheeler and Tan, 1983). For invalids are held to be in a delicate state, so this is matched literally *and* metaphorically by the light and careful cooking style prescribed.

More fundamentally than suiting the style of cooking to the occasion, people have decided views on whether food should be cooked at all, or eaten raw. Some British men think of salads not only as insubstantial, but inappropriate for a proper meal—'dinner isn't dinner unless it's something cooked'. In many societies, meat is virtually never eaten raw. The only well-known recipe in Western cuisine that calls for the use of raw red meat is steak tartare in which, significantly perhaps, the rawness of the meat is disguised by the method of preparation—buttock steak is scraped with the edge of a teaspoon to produce fibreless mush that is then further disguised by the incorporation of onion and egg. Even so, the dish is rarely mentioned in popular paperback cookbooks that are currently available in Britain—and the one well-established example that does include it advises the reader, if the idea horrifies, to pass on quickly (Innes, 1971). To us the images of eating meat in its raw state, dripping even only a little with blood, staining chin and hands, or of tearing flesh with the teeth, are associated with a notion of 'blood sucking' and conjure up monsters or some sort of semi-human being—the stuff of horror movies rather than orderly, civilized everyday living.

As Edmund Leach (1970, p. 34) has observed, the case of cooking could not be more apt:

> Our survival as men depends on our ingestion of food which is part of Nature; our survival as human beings depends on our use of social categories which are derived from cultural classification imposed on elements of Nature.... Food is an especially appropriate 'mediator' because when we eat we establish, in a literal sense, a direct identity between ourselves (Culture) and our food (Nature). Cooking is thus universally a means by which Nature is transformed into Culture, and categories of cooking are always peculiarly appropriate for use as symbols of differentiation.

Put another way, culturally prescribed practices for cooking answer the repeated but unspoken question about group identity. Literally and symbolically, we reassure ourselves, thereby that we *are* human not animal, civilized not savage; even though, like animals, we must eat to survive, we do so in a way that emphasizes that we are set apart from the animal world as a very special sort of being.

In this way we are 'explaining' our natural and social surroundings and making sense of our own place within them. The 'explanation' takes the form of a classification, a simple two-fold categorization of animals and people. This serves as a device whereby the paradox, that we are simultaneously animal and human, is resolved in favour of the latter, by our consuming of food in the very way animals cannot. As will be seen in the following three sections, this two-fold classification becomes more intricate as additional classifying criteria are added. Each section deals with a case of food avoidance. As we turn to consider the first—the case of vegetarianism—it is worth noting that, of all food avoidances, those pertaining to foods of animal origin, especially flesh, are accompanied by the most severe sanctions, engender the strongest reaction and are most frequently incorporated into religious observance.

'CRANKS'

A wholefood/vegetarian restaurant which opened in London more than 10 years ago called itself 'Cranks'. People who refuse to eat meat on principle or who choose whole flour, brown rice and avoid refined sugar have long had to suffer suspicion, even hostility; the 'red beans and sandals brigade' are still pilloried at medical conferences, the 'lentils and muesli belt of North London' sneered at by stand-up comics. The majority of the population of Britain eats meat and uses refined foods as a matter of course. To do so is 'normal'. People who do otherwise strike us as peculiar; all the more so, since to avoid meat and to prefer wholefoods is not, in industrialized society, widely associated with membership of a religion that prescribes vegetarianism. So it becomes puzzling; the only sense that can be made of it is to consider adherents to be faintly dotty.

Vegetarianism is clearly a logical response for those who believe killing animals to be inhumane or who see meat production as wasteful and ecologically destructive, but there also appear to be further and less obvious aspects of the preference for a vegetarian and/or wholefood diet which require some investigation. Twigg (1979, 1983) starts her account of vegetarianism by looking not at the significance of meat *avoidance*, but of meat *eating*.

Typically in our society we temper the character of meat, particularly its bloodiness, by cooking, and, as Twigg shows, the bloodiness or otherwise of meats is a feature of a readily recognizable hierarchy of foods. The basis of this hierarchy, to which vegetarian *and* non-vegetarian subscribe, is both status and some sort of power. At the top, highest in status, there stand the 'red' meats—steak, roast beef; lower down are the 'white' or 'bloodless' meats—chicken, fish; and below them, animal products such as eggs. The high-status items figure as the centre-piece of meals, though cheese and eggs are the centre of lighter meals, snacks or supper. On the next level down the hierarchy come the vegetables—regarded in our familiar meal system as adjuncts, insufficient to form a meal alone, without some sort of flesh or animal product as the focus.

An obvious feature of this hierarchy is the bloodiness or otherwise of the items. In general, blood is thought of, unsurprisingly, as an especially important substance. But it has also a strange emotional impact; the squeamish faint at the very sight of it; blood is 'tainted' to those who cling to a somewhat outmoded view of the inheritance of mental instability or of moral turpitude, 'noble' to those who likewise see virtue passed from one generation to another. Also, as Twigg reminds us, blood in meat is associated with strength, aggression, power and virility. It is still the practice in some parts of southern Spain for a woman to go to the butcher following a bull-fight to buy a steak from the bravest bull for her husband's supper, thus ensuring his continued strength and manliness. Red meat has long been associated with sexuality. Meat was seen as flesh and thus as carnal. In the traditional view, if virility was to be assured, red meat was prescribed; however, if an excess of sexuality was feared it was to be avoided. Twigg (1979, p. 20) cites examples of late nineteenth and early twentieth century books with titles like *Better Food for Boys* which offered vegetarian diets as the 'greatest aid we can give boys in the fight against self-abuse'! Red meat was not only sexual, it could also be overpowering to the sickly; Victorian and Edwardian invalid diets called for chicken or white fish. Invalids were thought to need delicate food that was not too rich. Roasting was considered too 'exciting' and 'lively' a form of cooking; instead food should be poached, or lightly boiled.

There are further features of this hierarchy with which the question of blood is related. Bound up with notions of strength, there are meats, red meats at that, which none the less are generally avoided. These are meats from uncastrated animals and from carnivorous animals. André Simon (1963) includes an entry in his encyclopaedia of gastronomy for lion, adding laconically 'a man-eating mammal not usually eaten by man but apparently fit to eat'. But in our own sort of society we avoid the consumption of carnivores, although other cultures do use meat from such sources. Twigg suggests that one reason for this avoidance might be that it involves a sort of double dose—to eat animals that eat animals becomes 'too strong'. Similarly, we tend not to eat uncastrated beasts; the meat from bulls and boars is likewise considered to be in some way tainted, too strong; remembering, of course, that there seems to be no nutritionally founded basis for such avoidance.

All of this serves as a reminder to those of us who eat meat that, even if we do so wholeheartedly and with enjoyment, and whether or not we view vegetarians with a slightly mystified tolerance, we too do not eat meat indiscriminately. Within the category of meat we make important and significant distinctions; we too engage in types of meat avoidance. Put the other way around, the sorts of discriminations that meat eaters make are exactly the same as those made by vegetarians. Vegetarian and non-vegetarian agree; vegetarians simply take it all rather further. For instance, vegetarians display evidence of adherence to the hierarchy already described—the *same* hierarchy that makes sense of meat eating. Typically, there is a process in becoming a fully fledged vegetarian whereby the first items to be given up are red meats, then poultry, followed by fish. This can continue logically into becoming a vegan and giving up animal products altogether. And it is not uncommon to find a self-styled vegetarian who cheerfully eats fish, and even chicken, on occasion, briefly returning 'up' the hierarchy.

Just as certain sorts of meats are deemed too strong either for everyday consumption or for those in a delicate state, so vegetarianism has a long history of association with abstinence, rejection of the body in favour of some sort of spiritual purity. Moreover, modern vegetarianism is associated with other sorts of social movements concerned with bodily and spiritual purity such as yoga, or naturism. As Twigg reports, vegetarians glorify bodily health and, at times, even interpret salvation in terms of it. The baser, animal aspects of human nature are less attractive, and ought not to be encouraged. These aspects, vegetarians maintain, are nourished by the consumption of meat; avoiding meat eating helps avoid the animal character of being human, and by the same token enhances the pure and more wholesome properties of people.

It becomes very clear, however, that a vegetarian conception of nature contains ambiguities and contradictions. Being vegetarian is explained and justified in terms of being 'natural'. But, on one hand, it is *natural* and true to essential human nature to avoid meat for it too obviously makes us one, as Twigg puts it, 'in substance and action with animal nature'; yet at the same time, vegetarians also reject eating meat because it is *unnatural* to do so. This time, unnaturalness is counted as a version of cannibalism; since, the argument runs, we as living beings are one with nature, then in consuming other living beings we eat ourselves.

Vegetarians resolve their ambivalence towards nature by a series of explanations. The bare bones of these explanations rely first on the assertion that, fundamentally, human nature is good, and so too is nature itself. So there is an emphasis on gardening, on bringing the rural into towns, on herbal cures to deal with bodily ailments—the organic is preferred to the scientific. Society becomes identified with urban living and thus with a falsity; it is artificial, distorted, essentially unnatural. The true expression of humanity lies in its pre-social aspect, the aspect at risk of distortion when confined by modern living. In this way, vegetarians regard avoiding meat as the expression of the naturalness of life and the essential virtue of human nature; vegetarianism becomes 'the natural diet of humankind'.

In his study of the symbolic significance of health foods, Atkinson (1980) finds expression of a set of beliefs about the damaging effects of society in general, and urban dwelling in particular, on the basic purity of human nature incorporated into explanations arguing for the value of such products. He finds that a very wide range of foodstuffs are promoted in a way that stresses their 'naturalness'. It is not only brown rice, sun-dried fruits or stone-ground flour that are presented in this way, but also the more commercially produced goods; 'Food from Nature: Yeast and Herbs: Bio-Strath', 'Potter's herbal and natural remedies can help you'. Time and again the naturalness of health foods is offered as an antidote to the unnaturalness and synthetic nature of modern living, and with it, of modern eating. This theme can be detected in a number of guises. For instance, health foods are grown with the help of organic rather than synthetic fertilizers, thus ensuring their naturalness. Wholefoods are essentially pure by virtue of being neither overrefined nor contaminated with artificial additives. What is more, their naturalness is attested in that they are traditional, even exotic, belonging to some older, wiser culture with 'a hint of Eastern promise'.

Since the publication of Atkinson's study, even the highly processed products found in supermarkets have been promoted more and more in terms of a rhetoric of traditional fare, of old-fashioned wholesomeness and natural goodness; at the time of writing one can walk into one British supermarket and buy for 45p a pound of rice in a plastic bag bearing the legend 'Natural Brown Rice'.

Putting this another way, food and eating are matters of morality. What people eat is an expression of a series of judgements about what is good for them, not just in a nutritional sense, but in terms of wider, often more diffuse beliefs about the nature of humanity itself. The Western version of vegetarianism is not associated in any direct way with Eastern religions. None the less, modern vegetarianism involves a documented set of beliefs that look remarkably religious. For it involves the expression of views, an ideology, about the relationship of humankind to the world in which we live, to other species in it, and about the essential nature of human beings. It indicates both the way, if not to damnation then at least to the degenerate (in this instance the evils of modernity), the path to be avoided and the path to 'salvation' are clearly marked (cf. Twigg, 1979, pp. 24-26).

As Twigg (1979, p. 14) summarizes it:

> Vegetarians make much play of the physical propensity of meat for corruption, and it is true that vegetables offer by comparison very little danger of food poisoning. But vegetarians then go on to regard meat literally as rotting matter, sometimes equating it directly with excrement. Through metaphorical transformations this rottenness as a quality of food becomes rottenness in other senses: the corruption of human relations or of the state for example. A physiological quality is thus taken up and made part of a much larger scheme concerning the nature of impurity generally.

Eating has, therefore, a moral and symbolic quality. When someone chooses to avoid eating flesh, when someone else consciously selects wholemeal scones and organically grown spinach, when a third prefers the pot of meat paste decorated with a picture of a man following a horse-drawn plough, they are making statements about themselves, answering the unspoken question 'Who am I?' by saying that they are a right-thinking sort of person with a concern for their health, the state of the world in which they want to live and the sort of qualities they regard as essentially human. As will be seen in the next two sections, the same story crops up over again in different ways.

DISGUST AND THE CAT

Some sorts of eating are universally considered to be particularly revolting. Meat avoidance crops up again, for most often it is members of the animal rather than the vegetable kingdom which seem to inspire the greatest degree of disgust. Literally dis-taste, disgust occurs in various guises.

In the 1930s a well-to-do German family followed the Christmas tradition of serving fresh carp. The fish was stored live until needed, using the bath as an aquarium. Come Christmas, however, their small boy refused to touch a mouthful of the carefully prepared dish, protesting that he could not possibly eat a fish to whom he had been introduced. The refusal to eat animals we take as pets seems to be especially powerful. The Nambikwara, mentioned earlier, may relish the louse from a relative's hair, but they live in intimate contact with a large variety of animals whom they fuss and fondle just like their children:

> They share their meals, are petted, deloused, played with and talked to, just as the children are. The Nambikwara have many domestic animals—dogs, cockerels, hens, monkeys, parrots, various kinds of birds and sometimes pigs and wild-cats. Only the dogs have any practical use. The women take them when they go hunting with sticks, but the men never use them for hunting with bows and arrows. The other animals are kept simply for pleasure. the Indians never eat them—not even the eggs which the hens go off and lay in the brushwood. (Lévi-Strauss, 1955, p. 61)

Unlike the middle class German child, however, the people in this tribe come perilously close to starvation for 7 months of the year. When manioc, which is the staple of their diet, is scarce, and hunting is simultaneously precarious, they will do no more than rely on whatever the women can collect. Even in such straitened circumstances, eating the pets is never even contemplated.

Eating the cat is unthinkable, but we do hear stories of such unthinkable eating. We may treat our own pets as special, but we are often fearful of what strangers might do to them. Cautionary tales continue to circulate in Britain about the dangers of eating a meal in a Chinese or Indian restaurant: where is the neighbour's cat? what about the dog that went missing? Strangers are suspected of doing terrible things. And even if foreigners are not accused of eating the cat, they are still held to indulge in disgusting habits. The Mukaranga are repelled by the nearby Shangaans, who eat tortoise, and other peoples in the Zambesi valley that eat lizards; and elsewhere in Africa, the Massa are horrified by the neighbouring Tupuri's weakness for vulture meat.

Accusations of this sort stick. The British disparagingly refer to Germans as 'Krauts', and Americans will call Italians 'Macaronis'. And this habit of naming strange people by the strange food they eat occurs in parallel fashion in cultures distant from our own. Dhor, which means 'eaters of beef'—otherwise forbidden—is the name given a subdivision of the Katkari caste of Bombay, and in the Congo the Western Lange, who eat dogmeat, are accordingly known as Baschilambua, i.e. dog-people, by the rest of Lange society.

If eating pets is unthinkable, eating people is insupportable. Cannibalism is universally the most disgusting form of eating imaginable. What is more, it is a standard accusation levelled at strangers. So horrified, for instance, was the Roman elite by stories of body eating that they set up special inquiries to discover what was involved when Christians took to holy bread and consecrated wine. Indeed, the American anthropologist W. Arens (1979) persuasively, if controversially, argues that cannibalism is no more than a myth, a widespread racist way of insulting and identifying strangers.

All this adds up to a three-fold association between strange people, strange eating and familiar animals. How do anthropologists make sense of it? One place to start is with the way peoples view their environment. All of us, American, Katkari and Tupuri alike, think about the world around us, its plants, peoples and so on, in an orderly way. Routinely, members of widely differing societies see their surroundings in very similar terms, by sorting their components into categories. We readily group things according to colour, size, familiarity—or, of course, scientifically derived taxonomies. As part of the environment, animals are just as likely as anything else to be subject to some type of categorization. One common basis for classifying animals is in terms of their closeness to ourselves and to our homes. We share our houses by choice with pets, our cats and dogs, or unwillingly with cockroaches, mice and other vermin. Dogs and cats, invited as pets, mice and woodlice, uninvited vermin, live alongside us, while lions and hyenas are very distant, in far-off countries or confined in zoos. Like other wild animals, they are alien and unreliable, dangerous, even sinister.

A systematic inspection of our own version of this classification, such as that undertaken by Edmund Leach (1964), reveals clearly that degrees of closeness distinguishes which animals are legitimately considered as food. Quite simply, we will not eat those which are either too close or too remote. Those we will eat are of two types: those we have domesticated and those we hunt as game. Cows, sheep and pigs are familiar, semi-tamed and docile, which we keep within our reach on the farm—within its boundaries but outside the front door of the house. Deer, pheasant or partridge occupy a similarly intermediate position between hearth and foreign parts as familiar inhabitants of our own countryside. The place of horses in the scheme is somewhat equivocal; at one time they used to work, they still provide entertainment and, most important, especially it would seem in Britain, they are also a sort of pet. At the farthest extreme, lion and rhinoceros are too mysterious and remote safely to contemplate as food. Right in our houses we may reluctantly find woodlice or even rats—too close, irksome and uninvited to think of eating. Pets, however, are welcome at home: dogs, so the English maxim has it, are after all 'man's best friends'. As such we invite them into our homes, pamper them, indulgently break the rules and share our food with them, spoil them and allow them up on the furniture. In short, we attribute human qualities to them wholesale. Effectively we elevate dogs to a sort of honorary human status—as indeed we may do with other species kept as pets. In this sense, people and dogs are the same sort of being. With an obvious reference to the taboo on cannibalism, dogs cannot be food. For we do not eat people whether they are real or honorary. To eat an honorary human being is to be less than human; to be less than human is monstrous and no better than an animal. By refusing to eat the cat we remind ourselves of our cultural identity, telling ourselves that we are the sort of right-minded people who would not dream of eating a friend.

Now the vehemence of expressions of disgust about strangers and their eating habits become clearer. As anthropologists and others have long pointed out, eating is one of the most intimate activities people can share. Who eats with whom indicates degrees of 'social distance' between people. 'Companion' is, after all, a word literally to describe a person with whom we eat bread; Judas's betrayal was especially terrible because he had shared the Last Supper. For eating together is a fundamental human social activity that enshrines and enacts awareness of common membership of one's own cultural group. Thus, you can only sit down to table with people like yourself. And by the same token you could not possibly share a meal with anyone who is likely to eat the cat.

BIBLICAL ABOMINATIONS

One aspect of food selection which, on the face of it, is easier to comprehend is that associated with religious observances. Without inquiring much further, a self-evidently acceptable explanation as to why some people avoid pork or others beef is to say that it is because they are Muslim or Hindu. In effect, however, this is not an answer, merely a restatement of the question. The puzzle still remains as to why Muslims came to prohibit using pigs as food, or Hindus cattle.

Close to our own cultural and religious heritage are the dietary rules of the Jews, and more particularly the extensive Old Testament list of clean and unclean animals that may and may not be eaten. Almost 20 years ago, the British social anthropologist Mary Douglas—who has since worked extensively on various aspects of the anthropology of food and eating (Douglas, 1974, 1975)—re-examined the abominations of Leviticus to try and unravel the question:

> Why should the camel, the hare and the rock badger be unclean? Why should some locusts, but not all, be unclean? Why should the frog be clean and the mouse and the hippopotamus unclean? What have chameleons, moles and crocodiles got in common that they should be listed together (Leviticus xi, 27)? (Douglas 1970, p. 54)

For centuries theologians have offered all sorts of explanations. These range from claims that the distinctions are no more than arbitrary, through ideas that they are a source of discipline which thus enables Jews to avoid thoughtless and unjust action, to the claim that the rules underlying distinctions between pure and impure animals are allegories of virtues and vices and serve as examples of good behaviour. Douglas, however, has no time for any of these. She dismisses them on the good scientific grounds that they do no better than proceed in a piecemeal fashion that fails comprehensively to account for the whole sweep of prescriptions and prohibitions. Pay attention to the texts themselves, she proposes, and note that every injunction is prefaced by the command to be holy. So it follows, she insists, that the rules must be explained in terms of that command.

As Douglas takes us through the complex variety of Old Testament meanings of what it is to be holy, she shows how classification of the natural world, and people's place in it, is once again central to the case, and this time adding classification and their relation to other-worldly affairs. We start off with the significance of holiness, and Douglas points to the way in which it means wholesomeness and completeness, in particular. Sacrificial animals must be without blemish, women must be purified after childbirth and so on. This extends to the social context—for instance, a task once started must not be left unfinished. And it extends still further so that categories and classifications by which the world is ordered must be tidily demarcated. One of these classifications is the ordering of animal species, subject, of course, to the same canons of neatness. So it follows that, for instance, hybrids are confusing and untidy, and thus to be abominated. 'Holiness requires that different classes of things shall not be confused. . . . Holiness means keeping distinct the categories of creation. It therefore involves correct definition, discrimination and order' (Douglas, 1970, p. 67).

Having got this far in her argument, Douglas is able to show that a comprehensive explanation for the basis of distinctions between acceptable and unacceptable foods is perfectly straightforward. She simply develops the notion of holiness on the same orderly lines. First she reminds us that the Israelites were pastoralists, herding cattle, camels, sheep and goats. Cloven-hoofed and cud-chewing, these animals provide a model of proper and usual kinds of food. This is endorsed, says Douglas, by the way that, if wild game is to be eaten, then it has to be of the same general sort as the food with which the Israelites were daily familiar. In other words, game animals were acceptable if they shared the distinctive characteristics of chewing the cud and having cloven hooves. In terms of these criteria, the hare and the hyrax (rock badger) are unclean: even though as ruminants each appears to chew the cud, their hooves are not cloven; and, vice versa, animals such as the camel, which are cloven-hoofed but not ruminant, have also to be excluded. Pigs, too, fall into this category. Note, says Douglas, that the only reason given in the Old Testament for avoiding pig is failure to conform to these two criteria.

> Nothing whatever is said about its dirty scavenging habits. As the pig does not yield milk, hide or wool, there is no other reason for keeping it except its flesh. And if the Israelites did not keep pig they would not be familiar with its habits. I suggest that originally the sole reason for its being counted as unclean is its failure as a wild boar to get into the antelope class and that in this it is on the same footing as the camel and the hyrax, exactly as is stated in the book. (Douglas, 1970, p. 69)

Having dealt with these prescriptions, the remaining prohibitions are easily explained. Another basis on which creatures are classified is according to the way they live in the three elements earth, air and water. Animals proper to each element are held in the Old Testament law to have defining characteristics. So in the air, to be real birds, creatures that fly with wings must also have two legs; in water fish swim with fins and are covered in scales; and on the earth four-legged animals either walk, hop or jump. 'Any class of creatures which is not equipped for the right kind of locomotion in its element is contrary to holiness' (Douglas, 1970, p. 70). From this scheme the classification of unclean animals can readily be derived: worms are on land but have no legs, insects fly but have too many legs, some locusts crawl and are thus prohibited, others hop and are thereby categorized as clean.

> If the proposed interpretation of the forbidden animal is correct, the dietary laws would have been like signs which at every turn inspired meditation on the oneness, purity and completeness of God. By rules of avoidance holiness was given a physical expression in every encounter with the animal kingdom and at every meal. (Douglas, p. 72)

The daily activity of eating thus embodies answers to the question of Israelite cultural identity.

Leviticus offers the anthropologist an especially convenient case. For not only are the rules for food selection codified as part of religious observances, they are written down and thus readily available for inspection. Obviously Douglas's account provides an analysis of Israelite avoidances. But it is also another example of the way exploration of food choice may proceed, whether or not the habits in question are explicitly linked with religious requirements. Attention to a people's classification of the natural world and to the place human beings have in it provides an explanatory 'template' that makes sense of habits which otherwise, from our own point of view, appear to have no sense. The next section tries to consider a set of food habits which, despite their contrived and commercial rationale, would none the less, from an adult viewpoint, make less sense without the lessons of anthropological analysis considered so far.

GETTING THEIR OWN BACK

'In a six-month period last year, 50 million ''Skull Crushers''—oozing red when bitten—found their way past the squeamishness of adults and into the mouths of the young in Britain, the EEC and Australia ... [and] are now bound for Canada and the United States.' This was how the *Observer* (27 February 1983) reported the manufacturing success of a newly established Scottish confectionery firm, and with it introduced a note of curiosity about what might lie behind ghoulishness on such an enterprising scale.

Much of the time children in most societies are in a socially subordinate position. They have little control over their lives; day-to-day decisions regarding their fate lie in the hands of adults, including, of course, plans for what, when and how they will eat. It is adults' ideas and arrangements that, by and large, are imposed on children. Parents decide when children are to graduate from milk to solid food—at 12 months old in Peru, 4 to 5 months in Belgium or Britain. And their progress in very many societies is marked by giving children particular food, held to be specially suitable for their years. As Wellin (1953) reports, bread in milk, and mashed potatoes in soup, are thought of as strictly 'baby' foods, and are appropriate only for Peruvian infants' transition from a milk to a solid food diet.

Control over access to food varies from one society to another. Now and again, however, children find scope for some autonomy. By the age of around 5, children everywhere become adept at making the most of their circumstances. If, universally, the food provided for children reflects each society's notions of what is suitable for them, so everywhere children get to an age where, actively, they start to forage for themselves. Children of one of the Australian aborigine tribes, the Wanindiljaugwa, are far more successful at wheedling titbits out of their fathers than are adult women. Children and youths regularly scout about and on occasion will by their own efforts secure a large prize, such as a stingray, which they will take with pride to senior members of the group, while lesser foods such as small birds and fish they usually eat themselves (Worsley, 1961). And it seems that in certain circumstances such gathered foods are nutritionally important, especially in those societies where the staple foods are often deficient in vitamins and minerals. Indeed, the diet of such children who forage out of necessity during a famine may actually improve (Farb and Armelagos, 1980).

This sort of foraging finds parallels in our own society as possibly unsanctioned and potentially risky raids on the fridge, or as excursions to the local shop to spend pocket money on sweets. This second version is, as hinted above, worth a great deal; reporting in 1982, the Trade Relations Manager for Trebor, a successful firm in this field, pointed out that 'The children's confectionery market certainly isn't kid's stuff. It amounts to some £1100 million of a total £2200 million confectionery market, and of this £1100 million about £530 million is accounted for by sugar confectionery' (Field, 1982, p. 4).

Undoubtedly manufacturers are successfully providing what children want to buy, and children are happily buying whatever the manufacturers have on offer. One look at any sweet counter in a corner shop or local newsagent's shows that there are some products which are exclusively for children's consumption. The manufacturers have done their market research and, like Peggy Field (1982, p. 4), regard children as 'shrewd, discerning buyers who know exactly what they want ... value for money, quantity, quality and novelty, though not necessarily in that order' (1982, p. 4).

Complementing this, James has done anthropological research on the sweet-eating habits of children aged about 11-13 which focuses on the way in which the kind of sweets children buy for themselves are markedly different from those that adults choose on their behalf (James, 1981).

In previous sections of this chapter, the concern has been with the consumption of the edible and the avoidance of the inedible. But when it comes to the case of children's sweets, we find the reverse: children seem to make a point of deliberately eating the inedible—metaphorically speaking. While engaged on research in North East England, James noticed that a local word 'ket' (used mainly in the area south of the Tyne) had two apparently quite unrelated meanings. When children used the word, they meant sweets—the sort they chose for themselves; but to adults, the word meant rubbish, anything worthless.

Taking a close look at children's ideas about sweets, James found a sharp contrast with those of adults. To adults, sweets are to be thought of as secondary to meals, nutritional extras. They round off a meal, or serve as a temporary stopgap if the next meal is delayed, but one has to ensure that they do not spoil the appetite for that meal. But for children, sweet eating is quite different. It is valued, enjoyed and sought after for its own sake. If anything, James suggests, 'It is meals which disrupt the eating of sweets'! As for kets, they are not sweets in the way that adults think of them. Indeed, in many ways they are manifestly not even food. Though technically as wholesome as any confectionery, kets differ from adults' sweets in ways which suggest *in*edibility: their name, their colour, the way they are consumed and so on. Thus types of ket have unexpectedly utilitarian names—Car parks, Telephones; others are of animals normally considered taboo—Jelly Gorillas or Lucky Black Cats. There are even hints of cannibalism—the well-known Jelly Babies, and newer Fun Faces, and Jelly Footballers—or the poisonous—sweet cigarettes. The colours are also not those of conventional foods; they are lurid, fluorescent blues, stripes or shocking pinks.

Above all, they differ in the way they are eaten. For adult foods there are 'table manners'. What is to be eaten must be kept free from bodily contamination, be properly wrapped and covered, and remain untouched until the time of consumption. It should not be fingered by anyone other than the person who is to eat it, and, of course, once food and sweets are put in the mouth, there they should stay. Eating kets, on the other hand, is a very different, often messy business. They may be stuffed, unwrapped and sticky, into pockets, then eaten unconcernedly in just the state they re-emerge; they can be shared by friends, offering a suck or a bite with no regard for hygiene and manners, and even kets which do not change colour in the eating may be taken in and out of the mouth to inspect progress. Such habits break all the adult rules for polite and proper eating—as children know only too well.

As James reminds us, children occupy a social sphere of their own, the playground, the streets and the corners, and the edges of adult conversation.

However, though this is distinct from the adult world, it remains dependent on it. Children may well make a world of their own, but they do so out of the bits and pieces the adult system provides. They reorganize the adult order, by the simple ruse of turning it upside down. In this way, James brings together the two meanings of the word ket. For metaphorically speaking, kets *are* rubbish in that children elect to eat what the adult world considers non-food. And by eating the inedible in this limited—and safe—way children are claiming autonomy by subverting adult rules of eating. When children choose to eat Mummies, Jelly Feet and Skulls, they are reminding us that they may know adult rules for orderly eating but are not yet wholly prepared to obey them.

Children may well have lives apart from adults with eating habits to match, but of course they also learn to share adult meals at home with the rest of the family. A family meal *par excellence* is the 'cooked dinner', to which we turn now.

HOME COOKING

The presumption that women will cook for men and children has an extremely long history and is found in culture after culture. Equally, people in all societies not only have clear ideas of what constitute edible and inedible items—i.e. what is to count as food—but also of what constitutes a 'proper' meal, distinguishing between meals suitable for special occasions and those for everyday. Preliminary work I have done in South Wales suggests that conceptions of a proper meal and beliefs about the appropriate social place of men and women come together as daily assurance of family identity and propriety (see Murcott, 1982, 1983a, 1983b).

It all revolves around what is known here and elsewhere in Britain as a 'cooked dinner'. Less a whole menu, it is a plateful of food consisting of the familiar 'meat and two veg'. This is more than a cliché, for it still carries considerable cultural significance. For the women in this study, a 'proper meal' was a cooked dinner; it is the meal which ensures the health and wellbeing of members of the household. It is the sort of meal to which people should come home after work or school. The women agreed on just what counted as a cooked dinner, and could readily specify the rules that it involved, both for the components and for the proper way of cooking and presenting them. Meat—flesh that was fresh, not offal or sausages—occupied the central place in the dinner. Chicken or turkey were acceptable substitutes, but fish was not. An essential accompaniment was potatoes, and so too was at least one further vegetable which had to be green, though there may be another one, two or even three different types. Finally, the whole meal was completed with gravy, poured at the last minute over all the items arranged on the plate.

The rules go further than this, to specify the selected cooking techniques that contribute to its being counted a proper dinner. First, the meat had to be roasted (baked) or grilled and vegetables must be boiled. Sundays were special—both meat and potatoes were roasted, and often the greatest number of accompanying vegetables were provided. There were alternative cooking styles but these had their drawbacks, were not quite 'right'. Chips instead of roast or boiled potatoes might be used, but this was cheating. Stewing meat and vegetables together in one pot in the oven definitely did not count as a cooked dinner, any more than did frying meat quickly, or serving raw salads. In other words, a real cooked dinner was food that demanded lengthy preparation and regular attention during the process of cooking.

Much the same sort of arrangements and requirements are reported by Fortes and Fortes (1936) about traditional Tallensi society. Fifty years ago, in what was then the Gold Coast, the standard meal was porridge and soup. 'Porridge', they say, 'is food, it makes you strong.' There can never be too much porridge, but 'the belly is spoilt' if too much of other things, like soup, is eaten. Porridge was never eaten on its own, but was accompanied by soup, which would contain a small flavouring of meat and was often made very pungent by red pepper and other seasonings. The soup was always put on to cook first. Each woman was responsible for the provision of meals for her husband and family. (Men may have had more than one wife, and on occasion co-wives cooperated). While it was boiling, the woman swept out the 'close' where the meal was to be eaten, and prepared the vessels for serving it. This took about 30-40 minutes. Then, when the soup was ready, she made the porridge, which was another 10 minutes' work. In all these tasks the woman was assisted by her young daughters. But, the authors note, while small boys sometimes lent a hand with the fire, they never helped with the cooking.

This Tallensi meal and the British cooked dinner both satisfy their respective cultural criteria of a proper meal, and 'meat and two veg' constitute 'food' equivalent to the Tallensi porridge. In South Wales all the household recognizes the cooked dinner as the meal which signifies their return home at the end of the day, thus underlining the change from employer's time to employee's, from school-time to home. It thus marks a significant interval in the day's routine. Equally, like Tallensi domestic organization, cooking in the main is the woman's work, whether or not she has a job outside the home, whether or not her husband also shares in the domestic work, and even, to some extent whether or not the husband is unemployed. For in the *ideal*, the convention persists that a husband's prime responsibility, is as a wage earner, to provide for his family, while the wife's is to take charge of running the home and above all to provide meals in the expected manner.

The cooked dinner effectively epitomizes the relative obligations of husband and wife—the former as breadwinner, the latter as home maker. The very nature of the cooked dinner, especially the preparation involved and the niceties of timing, prescribe when the wife has to be in her place in the kitchen to accomplish the task properly. A husband's job defines how he 'properly' spends his time during the working day in order to support his wife and children; by providing a proper dinner in the prescribed manner, the wife announces that she too has spent her time correspondingly honouring her obligations to husband and children. Cooked dinners are essentially family meals, provided by women for men and children; bachelors supposedly open tins, go without, or return to their mothers; the widowed, and housewives on their own, do not bother to 'cook' properly, if at all, for themselves; and husbands temporarily on their own revert to bachelordom. A proper dinner is taken at home, earned by the husband (father) and cooked by the wife (mother). It is proper to family life. Looking forward to going home to a dinner reassures that all is well at home, and that all are socially in their place.

CONCLUDING REMARKS

I have interpreted my brief to explore anthropological issues affecting food choice by focusing, on the whole, on food choices close to home and tried to indicate what anthropologists might make of them. I have deliberately avoided compiling a catalogue or formal literature review of anthropological contributions to the field, which means that the result is inevitably highly selective and no doubt idiosyncratic. (Contrasting lines of inquiry can be pursued in Goody, 1982; Farb and Armelagos, 1980; Jerome *et al.*, 1980; or Mintz 1979.)

All I have sought to do here is to say that, however unlikely it may at first seem, everyone, whether of a strange culture or our own familiar type of society, behaves and eats in a way amenable to an anthropological analysis that reveals an underlying rationale for apparently irrational or bizarre habits. Furthermore, we are doing more than we realize when simply living out our daily and familiar routines; when we sit down to the evening meal, come across unusual menus, catch the children with sticky fingers and bright purple tongues, or pet the cat, our behaviour reasserts the order of our world and our cultural identity within it.

ACKNOWLEDGEMENTS

The preparation for this chapter in its early stages was helped by conversations with Paul Atkinson and Sara Delamont, and latterly with Phil Strong who also made numerous valuable suggestions for redrafting. I am very grateful to them all.

REFERENCES

Arens, W. (1979) *The Man-Eating Myth: Anthropology and Anthropophagy*. Oxford University Press, New York.

Atkinson, Paul (1980) 'The symbolic significance of health foods'. In *Nutrition and Lifestyles*, ed. Michael Turner. Applied Science Publishers, London.

Bascom, William R. (1951) 'Yoruba food', *Africa,* **21**, 41-53.
Douglas, Mary (1970) *Purity and Danger: An Analysis of Concepts of Pollution and Taboo.* Penguin Books, Harmondsworth. (Pelican edition).
Douglas, Mary (1975) 'Deciphering a Meal'. In *Implicit Meanings.* Routledge and Kegan Paul, London.
Douglas, Mary, and Nicod, Michael (1974) 'Taking the biscuit: the structure of British meals', *New Society*, 19 December, **30** (637), 744-747.
Farb, Peter, and Armelagos, George (1980) *Consuming Passions: The Anthropology of Eating.* Houghton Mifflin, Boston.
Field, Peggy (1982) 'No kidding!' *Nutrition and Food Science,* **76**, May/June, 4.
Fortes, Meyer, and Fortes, S.L. (1936) 'Food in the domestic economy of the Tallensi', *Africa,* **9**, 237-276.
Gelfland, Michael (1971) *Diet and Tradition in an African Culture.* Livingstone, Edinburgh.
Goody, Jack (1982) *Cooking, Cuisine and Class.* Cambridge University Press, Cambridge.
Innes, Jocasta (1971) *The Pauper's Cookbook.* Penguin, Harmondsworth.
James, Alison (1981) 'Confections, concoctions and conceptions'. In *Popular Culture: Past and Present*, ed. B. Waites. Redgrave Publishing Company, New York.
Jerome, N., Kandel, R., and Pelto, G. (eds.) (1980) *Nutritional Anthropology: Contemporary Approaches to Diet and Culture.* Redgrave, New York.
Leach, Edmund (1964) 'Anthropological aspects of language: animal categories and verbal abuse'. In *New Directions on the Study of Language*, ed. Eric H. Lenneberg. MIT Press, Cambridge, Mass.
Leach, Edmund (1970) *Lévi-Strauss.* Collins, Glasgow
Lévi-Strauss, Claude (1955) 'I starved with the world's most primitive tribe'. *Realities*, September, 56-76.
Lévi-Strauss, Claude (1966) 'The culinary triangle'. *New Society*, 22 December, **166**, 937-940.
Mintz, Sidney W. (1979) 'Time, sugar and sweetness', *Marxist Perspectives*, Winter 1979/1980, 56-73.
Murcott, Anne (1982) 'On the social significance of the "cooked dinner" in South Wales', *Social Science Information,* **21** (4/5), 677-696.
Murcott, Anne (1983a) '"It's a pleasure to cook for him": food, mealtimes and gender in some South Wales households'. In *The Public and the Private,* ed. Eva Garmarnikow *et al.* Heinemann, London.
Murcott, Anne (1983b) 'Cooking and the cooked: a note on the domestic preparation of meals'. In *The Sociology of Food and Eating,* ed. Anne Murcott. Gower, Aldershot.
Simon, André L. (1963) *Guide to Good Food and Wines: A Concise Encyclopedia of Gastronomy Complete and Unabridged*, revised edn, originally published 1936. Collins, London and Glasgow.
Twigg, Julia (1979) 'Food for thought: purity and vegetarianism', *Religion,* **9**, Spring, 13-35.
Twigg, Julia (1983) 'Vegetarianism and the meanings of meat'. In *The Sociology of Food and Eating*, ed. Anne Murcott. Gower, Aldershot.
Wellin, Edward (1953) 'Child-feeding and food ideology in a Peruvian village'. World Health Organization. October. Mimeo.
Wheeler, Erica, and Tan, Swe Poh (1983) 'Food for equilibrium: the dietary principles and practice of Chinese families in London. In *The Sociology of Food and Eating*, ed. Anne Murcott. Gower, Aldershot.
Worsley, Peter (1961) 'The utilisation of natural food resources by an Australian Aboriginal tribe'. *Acta Ethnographica,* **10** (91-92), 153-190.

The Food Consumer
Edited by C. Ritson, L. Gofton and J. McKenzie

CHAPTER 7

The Rules Of The Table: Sociological Factors Influencing Food Choice

LESLIE GOFTON

Sociologists, unlike other social scientists, have had little to say about the food consumer. Most consumer behaviour texts contain sections on the influence of 'social factors' on consumption, and standard references are made to the influence of social class, or ethnicity, or gender, or age, on consumption patterns (see, for example, Reynolds and Wells, 1977; Zaltman and Wallendorf, 1979; Chisnall, 1975). Much of this work has been produced by market researchers, however; surprisingly little has been written by academic sociologists on the subject of consumption in general, and very little indeed on the subject of food (Kornhauser and Lazarsfeld, 1935; Murcott, 1983; Goody, 1982).

This silence is surprising. Sociology ranges over every aspect of the life of men in groups, and is concerned with the study of social institutions from the largest social structure, the organization of the state, through the institutions of political systems, work relations, the family, social class, and so on down to the level of the individual, and the way in which group life influences individual conduct, and forms individual identity and behaviour. Sociology insists that the human being is essentially and universally a social animal, and that central fact permeates every facet of the life that he leads.

Food is, of course, a live issue at the moment; hunger in the Third World jostles with the problems of agricultural policy in Europe and America for the headlines, alongside concern over the purity of food, and the dangers posed by the 'diseases of affluence'. None of these problems seems easy to solve; the lesson of much recent work has been that common sense is a poor guide to solutions. Famines, as Amartya Sen (1981a, 1981b) has shown, are not explicable simply in terms of failure to produce enough food, but occur as a result of changes in systems of allocation, often as a result of changes in entitlement.

The behaviour of consumers seems everywhere to violate the rules of common-sense rationality; famine victims may refuse to eat food sent as aid, consumers in the developed countries insist on following a diet which is bad for their health, while governments persist in policies which lead to the production of wasteful surpluses and taxation to meet the cost of subsidies.

Common sense again misleads us in our search for solutions or explanations for this behaviour. Barry Turner (1984, 1982) has recently argued that sociology has traditionally found difficulty in dealing with phenomena, such as sex and eating, considered to be founded on 'natural need'. They seem, despite their universality, to be capable of infinite variation, and this is a severe problem for a discipline which asserts the primacy of human invention and desire over natural need. What is the 'ontological' status of the natural? Where are the boundaries between natural need and human artifice?

Many analysts simply ignore these difficulties; given the fact of universal need, the explanation of human behaviour seems then to rest on the assumption of underlying rationality in social expenditures. 'Good' human nature, under this assumption, would deploy resources only to reach a modest level of comfort. Once needs have been met in this way, then any additional expenditure, unsanctioned by religious systems, is seen as the result of 'corrupt society' which 'tempts the individual to misuse material things, either for greed, or to serve envy and pride by rivalrous display' (Douglas, 1984, p. 5).

Western societies, of course, have become increasingly secularized. Traditionally, religion supplied the metaphysical framework, the values and meanings in terms of which consumption could be seen as having ends and purposes, both temporal and spiritual. The decline of such systems does not mean that consumers no longer imbue their actions with meaning, but the absence of these spiritual frameworks does present a problem for social analysts. As a consequence of this history, we tend to view human activity as falling into two main areas: those things spiritual, such as art, music and poetry, which are enjoyed "for their own sake", and those such as food and shelter which are pursued for 'base' physical instrumental purposes. Those which do not fall into the 'spiritual' category, and yet exceed the requirements of instrumentality are seen as making illicit, wasteful and unwarranted demands on the resources of society (Douglas, 1984, 1976).

When does need become desire? Modern consumption is seen as wasteful and immoral in a world ravaged by poverty and hunger—but clearly, this assumption rests on a particular notion of how such resources *ought* to be deployed, and ultimately on assumed limits of natural need. Analysis of the uses of food based on these assumptions totally fails to consider the social and historical dimensions of food consumption. The model of human agency upon which it rests is, effectively, desocialized.

Yet food consumption is a social matter. We do not just ingest nutrients, as Magnus Pyke has remarked; we eat food (cited in Brecken, 1975, p. 8). What we eat, how it is prepared, when it is eaten, with whom, what implements we use, how we behave before, during and after its consumption, the rules and meanings which permeate every aspect of food consumption practices are all social matters, irrespective of the biological, psychological, economic or political dimensions which they undoubtedly also possess.

This chapter will offer an account of some recent work in exploring that social dimension. It can only be a sketch of these aspects, since they are wide ranging. We begin with the idea of pollution; disgust is a universal phenomenon with regard to food, and Anne Murcott, in her chapter of this volume, considers work which anthropologists have done in relating its occurrence to the cultures to which they belong. Norbert Elias (1978) has looked at the development of the whole notion of civilized conduct in the history of European societies, including a detailed examination of food consumption practices. He argues that this must be seen as a new form of social control arising as a complement to political and economic changes, and their consequences for social institutions such as the family, work and the class system. He insists that these practices reflect the stratification of knowledge and cultural resources, and also embody principles of social control which rest on heightened individual awareness of the formation of identity in the deployment of such resources.

For Elias, then, understanding food consumption practices involves analysis of historical data—but he insists that historiography alone is insufficient. These data must be analysed in terms of some underlying set of theoretical principles which give them order. In the case of food consumption, we can only make sense of these practices if we see them as social institutions which relate to other aspects of the social system—as embodying, in other words, the social relationships which are important for the society in question. This is, of course, not all that they do, but they are carriers of meaning in the same way as language, marriage rules or principles of architecture.

Historical studies reveal that differences in food consumption practices are, of course, largely founded on inequalities of economic and political power. The study of diet in nineteenth-century Britain reveals the development of social institutions which mark the new social classes off from each other, and give meaning to their everyday actions and the ways in which their lives are ordered and understood. The onset of industrial urbanism as a way of life creates new social relationships, and these are reflected in the ways in which food is divided up and used in the society.

Apart from differences in the kinds of food which the social classes consume, however, the study of diet in this period also reveals the extent to which ideas about food, and beliefs about the properties and effects of food, also embody social values, and notions of morality and proper behaviour which reflect the relationships between different age and gender groups, as well as the social classes. Tastes in food, as with taste in general, are the mark of the man.

This is the subject of a massive study of French society published recently by Pierre Bourdieu (1985). Bourdieu argues that tastes, and the distinctions which they substantialize, are a cultural system which enacts the relationship between the social classes, and which expresses the world views which membership of these classes involves. Tastes in food must be seen as related to tastes in art, or popular culture, to the images of the body which the classes hold, and of the relationships between the sexes. Tastes are socially transmitted and, as such, are carriers of the order of things in general, and of the order of society in particular.

To insist on the importance of historical data is also to accept the necessity of accounting for change. This chapter ends by looking at the changes which have taken place in one area. Changes in food consumption seem widespread in Western society at the moment, and would provide an attractive topic in their own right (Angel and Hurdle, 1978). Rather than attempt such an overview, I close by reporting on research into changes in patterns of drinking in North East England because they seem to embody many of the issues which should interest the sociologist. Drinking is changing in North East England because the social system within which traditional patterns were formed is itself being transformed by the decline of industry and changes in the lifestyles of the population which are a consequence of this. With new relationships between the generations and the sexes come new patterns of consumption, and new meanings for those activities. This chapter argues that we can only understand these changes, or any patterns of food consumption, by placing them in their social context, with all the implications that this has for the modes of analysis employed.

MANNERS AND DISGUST

Let us begin with the idea, surely unexceptionable, that food consumption practices, and food uses, are a function of social differences, or distinctions. The German proverb, that Man is what he eats, can be interpreted in a number of ways. It was clearly always intended to have a moral message; common sense shows that what is eaten, or whether one eats it, how it is eaten, how much, and so on are all reflections and expressions of the place of the individual within his society, his social characteristics, and the relationships between the groups of which that society is composed, and of which the individual is a member.

Thus, if we look at the food rules which govern, for instance, the prohibition on eating certain kinds of meat in Jewish or Islamic society, these only make sense as part of a way of living which marks the individual off from the rest of the world in certain ways—the adherence to food rules symbolizes the values which members of the group hold to be central, and echoes, or at least resonates with, the classificatory judgements in the natural and social world which members of these groups make on the basis of their 'cosmology' or view of the order of the universe (Goody, 1982, Dumont, 1970, Douglas, 1966, 1976, Evans-Pritchard, 1940).

In the same way, the food rules of any other cultural or religious group are concerned with expressing the values and the social distinctions which are important within that society. It follows that understanding our own food consumption practices involves an investigation of the ways in which they also express such values.

This does not confine our investigation to food products themselves. Our food consumption practices do not just involve what sorts of food are eaten, but also who eats what, when it is eaten, the rules of etiquette, the nature of the occasions on which foods are consumed, and the symbolic meaning of food. A moment's thought tells us that we use such knowledge, consciously and unconsciously, every time we eat. There are rules about what is to count as suitable or acceptable food, both in general terms and on particular occasions. Some foods are 'for' breakfast, some for Christmas dinner, some for snacks, some for 'real' meals. We choose the food which we offer to our dinner guest with care, as we do the wine that we drink, or the condiments which we sprinkle on it. The order of dishes is set, how we use the silverware, and how we react to the dishes which are placed before us. On a common-sense level, the knowledge which we have enables us to manage the impressions which we make upon others—it is a communicative resource which we acquire during the socialization process.

Food consumption practices are social constructs; the system which they form cannot be understood apart from the ways in which it interrelates with other social institutions, in the process of historical development. At one and the same time, the system provides a communicative resource, a language, which both expresses the main themes and values of the society and enables individuals to pursue their individual projects and purposes. Every occasion of usage is, then, both a reaffirmation of a world view and a subtle modification of its shape as the individual interprets and restates it.

FOOD AND HISTORY

A review of the history of food, such as Tannahill (1973) offers, reveals the diversity of food habits, and the variety of things which have been counted as food, but at the same time reveals a universal predilection for rules; at any place, at any time in human history, there are ways of eating, and things which can be eaten which are distinctive—other people do it differently and this is the cause of disgust, horror, amazement or fascination. A common accusation levelled against an enemy, for instance, is that such people are anthropophagists—that is, that they eat each other, or people who offend them.

Assumptions about the eating habits of other people, in "less civilized" societies, or in the "more primitive" past, almost always involve imputations of savagery—lack of hygiene, being less 'fussy' about their food and so on. Yet this is clearly unsupportable if we look at the rules which these so-called primitives actually have (Arens, 1979).

What is interesting about such concerns is their universality. In all societies, ideas of pollution, dirt, and corresponding notions of hygiene, health and cleanliness seem to govern food consumption, with each society claiming its system as a practical, rational way of ensuring a safe and healthy diet. Danger or impurity are not seen as coming from the same sources, however, and ideas about pollution or impurity are certainly not necessarily related to what we (in Western, industrial societies) think of as hygiene. Even if we look back to our own recent past, we see that these have changed in a relatively short time. Gabriel Oak (in Thomas Hardy's *Far from the Madding Crowd*) is thought 'a nice, unparticular man' when he refuses a clean cup for his ale: 'No, not all (...in a reproving tone of considerateness)...I never fuss about dirt in its pure state, and when I know that sort it is' (Hardy, 1874).

Pollution concerns are the furniture of our daily life, of course—present but taken for granted and hence effectively invisible. What we eat, where, how, who with and so on are also part of what Erving Goffman (1971) called the 'territories of the self'. Control over them is the mark of a competent, fully accredited adult member of society. We expect food to be free from 'defilement' and to be 'fit for human consumption'. We thus regard teeth marks of unknown origin as defiling. We will normally only eat remnants of food left by close kin or friends, and are polluted by scraps, smears, drops or froth from food contacts anywhere outside the mouth, in the sight of others, at least. Contact with inappropriate items even extends to those things which we know to be 'clean' if they are otherwise suspect; see Rozin's findings on the fly-swatter experiment, (cited p. 100).

At the same time, there are clear rules about the kinds of food which can be served to members of the family, friends, guests, casual visitors or strangers, and about the combinations of food which are appropriate for specific meals (Nicod, 1979). These rules are highlighted by examples where they clearly do *not* apply: some are in abeyance in the case of children or non-competent adults, depending on context or occasion. Thus, in an old people's home, Jules Henry (1966) notes the way in which 'inmate' status (non-autonomous, non-responsible) is signified in an old man's acceptance of the offer of another inmate's scraps of food—acceptance of such polluted food signals exclusion from the world of normal, competent adults who must keep to the rules of 'civilized' behaviour—in public, at least.

In the 'total institution', control over the resources which are used to constitute the self, including food, is essential for the process of restructuring that self. This applies to the public school as well as the asylum, the concentration camp as well as the monastery or the old people's home (Goffman, 1968; Henry, 1966; Bettelheim, 1970).

What are the roots of this system? It is clearly a feature of particular kinds of society, and clearly also something which has developed over time, since there were different rules in the past.

One of the few theorists to give a central place to the question of food consumption practices as vehicles for the expression of social values is Norbert Elias. His study of the development of European society in the Middle Ages (Elias, 1978; see also Elias, 1983) takes as a central topic the development of the notion of 'civilized conduct'.

This underlies, he argues, the rise of secularized and centralized monarchies as forms of state government. Secular 'civilité' replaces the traditional authority of the Church, and involves internalized standards of behaviour which are disseminated from the model offered by the 'courtly' behaviour of the ruling class through the emerging medium of print. The essential feature of these new rules of conduct—whether they affect food taking, toilet, behaviour in public and so on—is that they rest on an increasing repugnance for the 'natural' or 'animal' characteristics of human behaviour. 'Natural functions' become sources of shame: accordingly, the animal origins of food have to be disguised along with the animal side of human behaviour.

Civilization, in a sense, is the process of developing 'standards of repugnance', internalizing new criteria in terms of which behaviour can be assessed and evaluated. In contrast to preceding societies, the civilizing process replaces the face-to-face oral tradition with an impersonal set of standards. These are disseminated *from* the centre out and rely on intense self-awareness for their effect. Social control comes to be based on internalized standards and values rather than *collective* judgement. Food plays an important role in this process.

Elias argues that a 'standard eating technique' existed in medieval society which corresponded to a 'very particular set of human relationships and structure of feeling' (Elias, 1978, vol. I, p. 101; see also Braudel, 1973) although with immense possibilities for individualized nuances and modifications. Nevertheless, there were essential continuities; those of highest rank are pre-eminent—they wash their hands first, and take first from the common dish. Despite fashion, and variations in the form of utensils, there are persistent threads in particular forms of meaning—the secular upper class, for instance, indulged in extraordinary luxury at table. Their wealth and rank were made visible by the opulence of their utensils and table decoration; but the social distinctions which this display signalled were not simply maintained because of the *inability* of other social classes to mount such meals, but rather because of a general acceptance of the significance of the social differences embodied in such displays. For those in other social classes, nothing else was *needed* when eating and departure from the conventional forms of food taking would be socially inappropriate, and hence unacceptable, rather than a practical impossibility.

Conduct when eating cannot be isolated, Elias argues; it is a segment of the totality of socially instilled forms of conduct, and its standards correspond to a very definite social structure. Medieval practices were very different from our own, and remained relatively unchanged for long periods. The things which we take for granted were only gradually accepted over a very long period of time; the social standard which lies behind the use of an implement such as the fork can only be understood in relation to the sensibilities, emotional life and rational consciousness of people eating together and is not, as we tend to think, 'natural' or self-evident. Think of the complex rules which we have about how such an implement should be used. These have only been developed and come into general usage over centuries (Braudel, 1973, pp. 136-141).

Sensibilities were wholly different in the Middle Ages. Those things which seem 'natural' to us—disgust with something which has been in contact with the mouth or hands of another, embarrassment at the sight or mention of bodily functions, shame when one's own functions are exposed to the gaze of others—all these were absent from the medieval sensibility which was conditioned to forms of relationships and conduct which to us seem uncivilized or, rather, pre-civilized.

It is the *raising* of these thresholds of shame and embarrassment which suddenly appears in the books on etiquette such as those of Erasmus (1530) and his contemporaries. The impact of such books indicates those areas in which tradition *can* no longer be a sufficient guide. Erasmus and those who come after him state precepts as to how people *ought* to behave in order to promulgate new social values and establish standards of 'civilité'.

A new way of thinking, and a new way of seeing oneself and others, are implied by the change. From this new way of thinking, we can deduce the 'attitudes of the soul', or the psychology of others from their appearance, dress, demeanour and behaviour. Not only do we begin to be more reflective about our appearance, and how we behave, but also a *common* consciousness of the standards which are expected exerts a controlling influence over us. This marks a new form of social integration, with increasing emphasis on correct standards of behaviour, modelled on how those in centres of political power conduct themselves (Elias, 1983, pp. 117-145).

Table manners begin to assume greater importance and standard eating techniques and uses of utensils at table gradually become established. However, 'Nothing in table manners is self evident', Elias asserts: they are not the product of a natural feeling of delicacy. The knife, fork and spoon are not natural or technical implements, 'necessary' in some absolute sense, so much as objects whose purposes, manner of use and function have been gradually defined over centuries of social intercourse.

Neither are these to be viewed as isolated processes; we can regard, for instance, 'civilité' in eating as parallel to changes in speech, with usage, forms of address and conventions of grammar being developed and, indeed, disseminated in just the same way, with the court setting standards which influence the rest of society. Clearly, these processes represent a general transformation of human feelings and attitudes. Changes in social relationships, and the forms of integration which must needs accompany them, necessitate behavioural changes, and concomitant forms of social control (see also Foucault, 1967, 1970, on the ways in which systems of thought in other areas of life change).

Table manners, forms of speech, the layout of buildings or the types of food and the ways in which it is eaten—all these, for Elias, embody the social relations in the society which has produced them.

Taking meat eating as an example, Elias notes that in the Middle Ages, before the change with which he is concerned, three basic types of behaviour could be found. Among the secular upper class, throughout Europe, meat consumption was extraordinarily high. What seem to us fantastic quantities of meat were consumed by certain groups (Elias, 1978, p. 118). Second, a totally different regime obtained in the monasteries; here, ascetic abstention prevails, not through shortage, but through self-denial. Gluttony is regarded as a sin, and there are specific injunctions against it. Amongst the clergy, we find strange redefinitions of what is to 'count' as meat in order to circumvent this prohibition; thus, the foetuses of rabbits, which are excluded from this category, form part of the provisions of most monasteries (Tannahill, 1973). The third group, the peasantry, have their consumption limited by shortage however; cattle are a limited and highly prized resource, and the meat which they produce is reserved for the highest table. This still shows in the language which is used to describe meat in Britain; the words for the live beasts (sheep and cattle) come from the Anglo-Saxon which would have been the tongue of the herders, whereas the words for cooked meats (mutton and beef) derive from the French of the Norman upper classes who would have consumed them at table. The same restrictions of availability apply also, of course, to the meat of wild ('game') animals. Such were the limits on the kinds of meat available that in Britain poaching 'was not only the livelihood of outlaws but the passion of men of all classes' (Trevelyan, 1979, pp. 22-23) and laws were passed to limit the keeping of sporting nets and dogs to those earning more than 40 shillings per year.

When meat is served, the rituals and behaviours involved reinforce the social distinctions which go with meat consumption in the medieval period. Amongst the upper classes, great importance is given to the presentation and display of the meat; sometimes the whole animal is brought to the table for carving, and, at the least, large parts of it are made the centre of the meal (what is called in England the 'joint' or the 'roast'). Carving itself is thought to be a necessary part of the social graces which every well-bred *man* should possess—carving is carried out by the male head of the household, and is regarded as a social honour, being 'handed on' to the male children when they reach the appropriate age (Flandrin, 1979).

The roots of this social institution, it has been argued, can be traced to the significance of meat eating in the social, political and religious life of the Greek city-states. Marcel Detienne and his colleagues at the Centre de Recherches

Comparées sur les Sociétés Anciennes have analysed the significance of meat sharing for the Greeks, and argue that this functions 'to enact and re-enact a number of social, political and sexual relations, to define Greek culture as opposed to other cultures, to express man's relation to the Gods—who do not eat meat but only inhale the smoke from sacrifices—and finally, to distinguish man from animals, eaters of raw meat, not cooked' (Svenbro, 1984). They go on to argue that the ideas of carving, division of the carcase and sacrificial butchery form a paradigmatic set of principles which underlie Greek thought at all levels, from politics and religion through to science and philosophy (Detienne, 1977).

The important point for Elias is that meat eating similarly "substantialises" social relationships within medieval European society. Meat is a tangible expression of social rank and prestige, and the family, which is still a unit of production, is expressing both the social relationships which it has to the rest of society, and the social roles within it, in the practices which it follows.

This ritual significance declined with the 'waning of the Middle Ages'. In France, Elias notes that by the seventeenth century, carving had ceased to be a necessary social grace, a mark of 'breeding' among the upper classes, and shows the same decline in importance in other societies. This change in behaviour and social values reflects, of course, other social changes which have been taking place, the most important of which are changes in the nature of the family.

Family units have become smaller during this time, and households themselves have shrunk from the form which they had during the Middle Ages. Changes in the economic organization of society have removed production and processing (for instance, weaving, spinning and other craftwork, but also much of food production and processing, including such things as slaughtering and brewing) from the home, and taken them to specialized premises, and these have more and more been given into the hands of specialist workers. From being a unit of production, the household has become a unit of consumption; to an ever-increasing degree, food consumption practices reflect this separation. Without suggesting that this change takes place overnight, Elias argues that this change in the forms of experience which make up the everyday life of the household gradually finds a natural corollary in a change of sensibility. Dead animals, or food in a form which makes the nature of its raw material too apparent when it is presented at table, become increasingly repugnant. Elias argues that we see a clear movement from a sensibility which views the sight and the carving of a dead animal at table as a pleasurable, or at least normal, experience to one in which any reminders that 'meat' has to do with 'dead animals' are to be avoided. Many meat dishes have assumed forms so far removed from their origins that the connection is almost totally obscured.

The civilizing process involves the development of social standards by which societies seek to suppress in the behaviour of their members every characteristic which they feel to be 'animal'. It follows that the animal nature of food should also be concealed—hence the 'social pre-digestion' of food, and the transformations of cooking (Lévi-Strauss, 1970). Clearly, the removal of the joint from the dining table and the concealment of carving have not taken place to the same degree everywhere; in England, the joint still survives, but during the nineteenth century the adoption of the 'Russian system' of table manners relegated it to the side table in 'polite' society. Often carving is removed to specialized enclaves behind the scenes at mealtime.

Other general themes can be traced by looking at the development of the rules governing the use of eating implements and plate changing. The practices which participants follow, and the ways in which they respond to the framework which this rule structure provided, involve the same kind of communicative action as other areas of social life, including language. In just the same way as language, this system is not to be regarded as fixed, given or invariant. Social order is not simply a set of pre-existent rules: it is incarnate in social practices of all kinds, and as such is constituted and reaffirmed, but is also constantly reinterpreted and reformulated in the very processes which give it substance.

Elias argues, then, that in the area of food consumption we can trace connections between the form of social order in a society and the rules for food consumption which are followed in that society. The idea of something like table manners, building design or modes of speech embodying or substantializing social relationships is interesting, but of course it is not particularly novel. This is a notion which archaeologists have pursued for many years. We can find, in the physical form, the manner of use, and so on a great deal about the social relationships and forms of behaviour which were present in ancient societies.

Elias adopts a very broad historiographic method, but insists on 'hard' evidence; historical analysis can only be effective, however, if its data are understood in terms of some principles of theoretical order.

INDUSTRIALISM AND FOOD

The history of diet in nineteenth-century Britain reveals that, during this period, food consumption patterns were being directly influenced by the great social changes which industrialization, and the rise of urban life, were bringing.

John Burnett's excellent *Plenty and Want* (1966) chronicles the significant features of food consumption during this period, and brings out the clear social divisions which are signalled by food. The diet of the poor is marked by shortage and economic exigency; home baking and brewing are all but wiped out by the lack of fuel and space in the new industrial towns, while food quality declines with the largely unregulated rise of commercial production, distribution and retailing. The dominance of the middle and upper classes is signalled by unprecedented indulgence, with a diet composed of huge quantities of the finest food. Social mobility comes to be signalled by the adoption of French cuisine, and the structure of the meal itself moves away from the traditional

English 'all-in' course, to become a succession of dishes in two great courses, divided by 'entrées' in the first course, and 'entremets' in the second (Burnett, 1966, p. 58). This is accompanied by the adoption of French and German manners of serving (sending the servants around with the dishes) and the appearance of cookery books which codify and disseminate the new standards. The richest country in the world attracts the finest cooks, but their influence is not felt only by the upper classes but also by the aspiring middle classes, who accept the model of 'good living' which these habits embody. Even the clerks and shopkeepers could afford to take on a little of this grand manner at a time when even a modest income might support one or two servants. For the poor, cookery books written by the great chefs provide instruction on eating well on a limited budget.

For the rich, this also marks the beginning of a growth in entertaining, not only for display, but also as part of commerce. The gentlemen's club is as much a product of new patterns of residence, with business being done in the city while the family home is in the suburbs, as it is a mark of social distinction. Mealtimes are also changed by the new patterns of work which industrialism brings to all social classes (Palmer, 1952).

For the mass of the population, however, the nineteenth century was not one in which their living standards or their diet improved greatly. Contrary to popular myth, this time was not for them marked by widespread availability of new products and the growth of a new consumerism. Rather, this is a period marked by social evils; an exploited group finds itself at the mercy of commercial interests which are suddenly unfettered by the prevailing commitment to *laissez faire* and free market ideology. Among the first things to go are the centuries-old statutes controlling the price of bread and basic foodstuffs. This is the era of adulteration, short weight and the company store.

A social history of food in nineteenth-century Britain is a disconfirmation of mythology as much as anything else. Nineteenth-century 'progress' consists of new ways of showing social distinctions rather than a general improvement in the standard of life enjoyed by the general populace. This is certainly very evident in their patterns of food consumption.

A great deal of reforming zeal and illustrative research was concentrated on the diet of the poor, and produced horrific accounts of the deprivation which they suffered. The expression of social distinctions in terms of dietary differences is all too apparent here; during this period the language of social differences is given a grammar and a vocabulary which have survived, to a large extent, into the present.

We can supply the dimensions of this language easily enough; the names and times of meals still mark differences between the social classes in the United Kingdom (Palmer, 1952). Dinner, for the middle classes, moves into the mid-evening in the early nineteenth century, while for the working classes it becomes a midday meal; lunch (for the gentry) is a light snack, taken in the saddle, or sur l'herbe, as often as not.

The times of the meals differentiate between the classes but, of course, the forms of food, and the ways in which the meals are eaten, also mark social differences. The diets of the classes are marked not just by the absolute differences in the amounts of food consumed—which are considerable—but also, even when we look at examples of working-class diet in which poverty does not operate as a critical constraint, by the types of food which are preferred and the ways in which they are used.

Amongst the working class, we see a diet based on bread, poor types of meat and a limited number of vegetables; this is largely dictated by economics, of course, with considerable differences between the regions, and between town and country, but it becomes a 'preferred necessity' for many members of this class.

The diet of the upper classes is a self-conscious display of power, excess and ostentatious wastage. Important elements in this are those game birds and fish which the laws passed in the late eighteenth and early nineteenth centuries reserve even more strictly for the sport and table of the gentry—contrary to all the history and tradition of the countryside. The nineteenth century marks what Harry Hopkins (1985) has called the 'poaching wars' with the enclosure of common land to create gentlemen out of the newly rich industrial elite, and the enforcement of harsh regimes of repression against the crime of poaching. The mountains of beast and bird slaughtered for feasts which are far beyond the digestive capacities of the guests, contrast starkly with the scale of hunger in the rural population which depends on the price of grain and the demand for agricultural labour.

The sheer contrast in the diets of the social classes illustrates how immense are the social divides; among the poor, adulteration is so widespread that the Cooperative Society hires a lecturer to tour the country giving advice as to what food should really taste like (cited in Burnett, 1966, p. 202). Those who have endured the horrors of the 'apprentice system', in which orphans are boarded out as 'apprentices' to a factory owner, describe how they fought with pigs for swill, since it was better than the food which they were given.

And yet, we have this image of the nineteenth century as filled with Pickwickian Christmases and opulence. Even everyday meals, for the upper classes at least, were staggering: 'Not since Imperial Rome', J.B. Priestley remarked, 'can there have been so many signposts to gluttony.' Groaning sideboards at the breakfast table, eight-course lunches, anything up to twelve courses at dinner, and a sumptuous tea in between. Meat was the central feature of this gluttony, and the shooting parties of this period somehow express the nature of the Victorian and Edwardian upper classes extremely well. The land has been appropriated and the rights of gentlemen to the birds and beasts on it secured by parliamentary Act. The railways and the roads are built which make it easy to assemble parties at distant estates for the weekend's sport. Improvements in guns (breech loading, smokeless powder) make them easier to shoot. Bird rearing itself becomes 'industrialized' with huge numbers of semi-tame birds reared and driven toward the guns by armies of beaters. The shooting party, in J.G. Ruffer's phrase, combined 'the opportunities of a Vimy ridge machine gunner with an infinitely better lunch'. A party led by King Edward VII accounted for 4000 birds in one day; His Majesty accounted for one-quarter of the bag himself.

The diet of the English upper classes in the nineteenth and early twentieth centuries clearly embodies many aspects of the distinctions between the social classes. Their food itself has a ritual exclusivity; it is *reserved* for the upper classes not just by social convention but by law. Only they are entitled to the birds and beasts which form the ceremonial centres of their feasts. These feasts are orgies of conspicuous waste on every level. They absorb food in absurdly excessive amounts; meals virtually fill the whole day; and they involve massive preparation, with huge numbers of servants involved in cooking and serving at the homes of the wealthy. This clearly makes no sense nutritionally, and the diseases of affluence with which we are now familiar become widespread in this class at this time. We can only make sense of this, surely, by relating it to the other ways in which social distinctions were expressed; indeed, by seeing the whole system of social distinction as composed of interrelated elements.

Thus, if we turn to ideas common at this time on the subject of the influence of food on health, we can see that food is seen as a form of control, expressing the place of the individual in society and also in itself embodying ideas of restraint. For children, certain foods are forbidden because they are too rich or 'morally' dangerous in some way. The diet of children, John Locke averred, should be 'plain and simple. Flesh once a day, and of one sort at a meal, is enough . . . without other sauce than hunger, is best'. Bread is to be eaten in quantity, along with 'other things that we are wont to make in England' since these are 'very fit for children'. Sweeteners, spices and things that 'heat the blood' (or make the food taste more pleasant) are to be avoided. Any drinking between meals, and especially when hot, was to be strictly avoided. Regular bowel movements were absolutely vital, and with a highly farinaceous diet, Locke could simultaneously warn against the administration of 'purges or physics'. Other writers warned against the dangers of meat, rich food and high flavouring in the tracts on 'purity' which began to appear in the late Victorian period. These were claimed to 'inflame' the passions and promote self-abuse!

Those from other countries who did not follow this régime were viewed with horror. Mary Lundie Duncan wrote of America in the 1840s, that the children 'breakfasted on fish, flesh and game, fruits, salads and hominy; johny cakes, corn cakes, buckwheat cakes all hot with molasses, toast swimming in butter...tea and coffee'. This, she felt sure, would ruin their digestion. She was even more horrified by the fact that they sat up to dinner (whereas in England they would still be supping bread and milk in the nursery), sharing 'adult foods' with their parents. This, she noticed, seemed to 'cause' precocity and lack of childishness (cited in Avery, 1984).

As Gillian Avery notes, these beliefs and practices were largely unchanged in the 1930s, in middle-class households and in public schools. Meals were seen as a form of discipline, with self-denial and self-control as the watchwords. The most substantial changes could be seen in the attitude to the bowels, with a hysterical demand for regularity, constant recourse to syrup of figs and a propensity to blame most things on constipation. Control over drinking apart from meals was so strong that one public school turned all taps off during the day—with the result that some of the boys went off in search of water butts and puddles. Discipline also applied to the preparation of meals; anything which required too little preparation was somehow regarded as reprehensible. The diet of children, and indeed the diet of the English middle classes in general in the prewar period, is characterized by Avery as 'using few ingredients, and making the worst possible use of them'. Vegetables, when they were used, tended to be over prepared. The middle classes, Avery notes, placed great emphasis on 'effort' at the same time as they distrusted sensuousness and lack of self-discipline. Food for children (and in a different way for adults) served at one and the same time, as a vehicle for values of self-denial and self-discipline, and an expression of fastidiousness and effort.

> To give just raw fruit or cheese as a second course would be slovenly. Shop cakes or biscuits were a sign that the mother was not a properly caring one. Vegetable soups were painstakingly sieved through hair sieves (this went for babyfood too). Food out of tins was only for those who lived in bedsitters. To be seen to take short cuts was to lose caste. The preservation of your caste in the elaborate middle class structure of pre-war Britain was a highly important matter. (Avery, 1984)

Historical examples illustrate some of the issues which have been important to past societies, and enable us to make connections with the present. We now consider how this form of analysis can be applied to contemporary society.

A SOCIOLOGY OF TASTE

One of the most important works on the sociology of consumption is Pierre Bourdieu's massive study of the social significance of taste, part of which involves considering how food consumption fits into French society (Bourdieu, 1985; see also Lévi-Strauss, 1978; Barthes, 1979). Bourdieu's conclusions are uncompromising; for him, food consumption, as with the cultural consumption of all other resources, is a vehicle for social differentiation, an embodiment of class inequality and the stratification of knowledge, aesthetic sensibility and values.

His is a massive project: to relate, in a sociological analysis, the whole system by means of which different types of consumers, and different types of tastes for all cultural goods (from art to food flavours), are produced. The prevailing orthodoxy, the 'ideology of charisma', sees taste as a gift of nature; it is, Bourdieu insists, a product of upbringing and education—surveys of the most elementary kind find direct relationships between visits to museums, or foreign restaurants, and the level of education received, or parental income and home background. Tastes, he says, are 'predisposed' to function as markers of class. The manner of using a particular taste—for instance, being able to 'appreciate' food and wine, or 'read the meaning' of a painting or piece of music—refers to the manner of its acquisition as it distinguishes the individual who possesses it in specific ways (as someone who has invested time in developing the skills involved, who is able to use the words in the correct ways). This point is also made by Douglas and Isherwood (1979) and by David Reisman (Reisman, Glazer and Denney, 1950). Tastes are cultural capital which is deployed in communication exchanges (what Reisman has called 'taste exchanges'). This is built up by the investment of time in education, or direct experience, or the practice of particular skills or arts. The way of seeing (*voir*) which is involved is a function of knowledge (*savoir*) employed (in the concepts or the words which name the components of the act) whether it be describing the taste of a wine, the meaning of a sculpture, or the history of a football team.

These are, Bourdieu says, programmes for perception. A person who possesses the 'code', or the cultural competence to use words appropriately and legitimately, thereby has his experience shaped by it—he moves from the primary level of ordinary experience to the 'stratum of secondary meanings'. The pure gaze or the 'brute experience' is, Bourdieu insists, an invention; the 'eye', the ability to 'see' and to understand, is a historical phenomenon, which is only available through education, whether in the home or in institutions of learning (see also Berger, 1970).

Tastes in food, Bourdieu insists, are the strongest, most indelible mark of infant learning; the 'native world' of the individual is, after all, the world of primordial tastes and basic foods, and archetypal relation to the archetypal cultural good. At the same time, of course, foods preferred are 'indicative modes of self presentation' (Bourdieu, 1985, p. 79); they show off a lifestyle, along with other tastes, and those most crucially aware of this are, according to his research statistics, those most concerned with their economic, social and cultural 'trajectories' (i.e. social climbing)—that is to say, the 'petite bourgeoisie'.

Pointing to the unity between the purest pleasures of the aesthete (all of which are based on the most fundamental oppositions in categories—sweet/bitter, hot/cold, coarse/delicate and so on), these oppositions, he says, underlie class differences in taste; there is not a simple effect of income which can account for taste differences, since this would not accommodate situations where tastes are different although incomes may be the same. The real principle of preferences is based on taste. As one goes up the social hierarchy, so the proportion of income spent on food diminishes; the proportion of the total expenditure which goes on heavy, fatty, fattening foods, which are usually also cheaper (pasta, potatoes, beans, pork), declines, whereas an increasing proportion is spent on lighter, non-fattening foods. Now this appears to support the income hypothesis, but in fact it is an illusion. What we have instead is the operation of the whole range of social conditions which affect consumption tastes.

> The true basis of the differences found in the area of consumption, and far beyond it, is the opposition between the tastes of luxury (or freedom) and the tastes of necessity. The former are the tastes of individuals who see the products as material conditions of existence defined by distance from necessity by the freedom of facilities stemming from possession of capital; the latter express, precisely in their adjustment, the necessities of which they are the product. (Bourdieu, 1985, p. 177)

Working-class tastes emanate from the practicality of meeting specific objectives which are structured by experience, within limited means. Bourgeois tastes (the idea itself is premised on freedom to choose in some absolute sense) proceed from the idea that expressivity involves, at one and the same time, emancipation from necessity, and parity between those who do make choices—so that 'coarseness' in tastes is seen as a mark of inferiority in the individual rather than a consequence of social circumstance. The point, says Bourdieu, for the working class, is that they develop a 'taste for that which they are anyway condemned to' either as a virtuous necessity (thrift) or as a pathological preference for bare essentials ('a sort of congenital coarseness'). They wear their (lack of) tastes like stigmata, their lifestyle betraying their lack of cultural capital in every nuance; they exhibit their social position nowhere more strongly than in their approach to food; while the rest of the social classes seek to become slim and stay sober, Bourdieu's peasants and industrial workers explicitly challenge the legitimate 'art' of living, 'maintaining an ethic of convivial indulgence' (Bourdieu, 1985, p. 180) which is evidenced both in their attitudes, and in how they spend their money on food.

Bourdieu argues that these tastes tie in once again with the most fundamental attitudes; a modesty in entertaining and eating which is the mark of upper social classes speaks, he argues, of a discipline and deferment of gratifications which is a central aspect of a Benthamite calculation of costs and benefits. The sobriety of the bourgeois, he asserts, is affirmation of the attempt to escape from the necessity of the present, the immediate; the hedonism of the working class is an acceptance of the force of present circumstances.

At the same time, specific tastes in food mark specific cultural values. Foods which demand heavy investment in preparation time are significant expressions of a complete conception of the domestic economy and the division of labour between the sexes, and of the traditional woman's role. In the case of upper-class women, since their labour has a high market value, they tend to devote their spare time rather to childcare and the transmission of cultural capital and thereby to contest the traditional division of domestic labour. 'The aim of saving time and labour in preparation combines with the search for light and low calorie products and points towards grilled meat and fish, raw vegetables etc.' (Bourdieu, 1985, p. 184). In contrast, the working class Frenchwoman is tied to the preparation of dishes such as *pot au feu* to such a degree that the name of the dish has come to symbolize that conception of the housewifely role, just as the slippers symbolize that of her mate.

Food cannot be looked at in isolation, of course. For Bourdieu, there are three styles of distinction, or structures of consumption. Food is one; culture, in the everyday sense of 'art', is another; and presentation (clothing, cosmetics, furniture and so on) is the third. Clearly, these are related to each other. Thus, for example, tastes in food also depend on the ideas which each class has of the body, and of the effects of food on the body. These may be completely different for various classes. The working classes, he suggests, tend to emphasize the strength of the male body rather than its shape; this is reversed for the bourgeois (see also Fischler, 1984). What is important is that the conception of the body implies a whole regime of treating it, a whole way of caring for it and so on. Thus, in a direct sense, body properties are class-based. But also specific tastes in food are implied by the body scheme which is subscribed to:

> The physical approach to the act of eating governs the selection of certain foods...for example, in the (French) working classes, fish tends to be regarded as an unsuitable food for men, not only because it is a light food, insufficiently filling, which would only be cooked for 'health' reasons, i.e. for invalids or children, but also because (like fruit except bananas) it is one of the fiddly things which a man's hands cannot cope with, and which make him child like...above all it is because fish has to be eaten in a way which totally contradicts the masculine way of eating—that is with restraint, in small mouthfuls, chewed gently with the front of the mouth, on the tips of the teeth (because of bones)...the whole masculine identity—what is called virility is involved in these two ways of eating, nibbling and picking, as befits a woman or with whole hearted male gulps.' (Bourdieu, 1985, pp. 190-191)

At some length, Bourdieu develops the distinctions which are involved, for the French, in ways of using and referring to the mouth idiomatically. He argues, in much the same way as Elias, that such matters are deep expressions of the fundamental values of the social group, and that ways of walking, of talking, expressions and modes of behaviour encode social distinctions, deep values and fundamental beliefs; in a sense, this is a 'practical philosophy of the body' (p. 192). Clearly, this is the arena which a 'practical philosophy' of each class is enacted (p. 192). This is not reducible either to simple differences in goods consumed, nor even to the words which the classes use to describe what things are important to them; the first takes no account of quality, nor mode of consumption, while the latter ignores the different meanings which ostensibly the same concepts may have for different classes (think of what 'healthy food' means, at different times and for different groups). Rather, the differences arise from fundamental oppositions resulting from the social circumstances and values of the class groups concerned. The working class meal is structured by functional necessity and a concomitant emphasis on the substance and properties of food—the preparation, the division, the consumption of that food is centrally ordered around notions of practicality, necessity and substance, all of which are dictated by this class's experience and social position. The bourgeois meal, in contrast, emphasizes form, ritual, appearance and manners—the bourgeois denies necessity, makes a virtue of presentation, gesture and display and emphasizes the way in which food signifies social position and relates to other presentational features (bodily shape, furnishing, as cultural capital and so on).

Bourdieu is a committed social analyst who leaves no doubt as to his position, and yet he emphasizes that the moral implications of these 'embodied world views' cannot be viewed from a neutral standpoint. What is 'unpretentious' for the working class is 'slovenly' for the bourgeois; familiarity is a compliment for one class, and an undue liberty for another.

This mode of analysis is new and exciting; Bourdieu, for perhaps the first time in this area, has combined imaginative theory with interpretative and substantive survey material. Certainly, this raises some fascinating questions. How far can we go in tracing this fundamental set of principles in the behavioural and attitudinal pattern which we find in modern society? The symbolic form of analysis has always argued for this project, but it is only now being married to substantive empirical research .

FOOD AND SOCIAL CHANGE

To emphasize the connections between food use and social relationships depends, to some extent, on the assumption of stability, order and continuity in society. Such assumptions are dangerous. Societies are always, to a greater or lesser degree, in the grip of change, and the nature of such forces should be the urgent concern of any social analyst. These are clearly vital concerns in the area of diet and cuisine.

To talk of a sociology of the food consumer we must consider the ways in which diet is changing in Western society, and try to relate this to the issues already raised.

The semiologist Roland Barthes has written a good deal about the symbolism of food. In modern societies, he argues, food is used as a multi-leveled symbol, in that it

> serves as a sign not only for themes, but also for situations; and this, all told, means for a way of life that is emphasized, much more than expressed by it. To eat is a behaviour that develops beyond its own ends, replacing, summing up and signalizing other behaviour that develops beyond its own ends, and it is precisely for those reasons that it is a sign. What are those other behaviours? Today, we

> might say all of them—activity, work, sports, effort, leisure, celebration—every one of these situations is expressed through food. We might almost say that 'polysemia' of food *characterizes* modernities. (Barthes, 1979)

In the past, only festive occasions were signalized by food in any positive, organized manner. But today, work has also its own kind of food (on the level of a sign, that is): energy-giving and light food is experienced as a very sign of, rather than only a help towards, participation in modern life (Barthes, 1979).

Barthes sees modern usage of food as changing from that which fits into a ritual, an atmosphere or a mood, into that for which it is supposed to stand—the more situations develop a specific associated food, the more we are seeing the nutritive or natural function of food being suborned by its value as 'protocol'. In Barthes's view, contemporary French cuisine involves the natural outcome of this, that 'Food has a constant tendency to transform itself into situation' (Barthes, 1979).

The outcome, more generally, is that we are seeing very different relationships between food and the social situations in which it is used. The advent of mass-produced and/or fabricated food, convenient at home and 'fast' when eaten in the dinette, the increasing tendency to eat casually, the decline of formal meals, and the collapse of the social groups which used to eat together, mean that food habits are becoming increasingly individualistic, or anomic.

In an interesting aside, Sidney Mintz (1982) argues that what this amounts to is an inversion of the relationship between food and the situations which, traditionally, it could be said to signify. Food availabilities in modern society have tended, he argues, to 'smooth out' or to eliminate the structure of meals and the calendar of diet in daily life; modern foods, such as soft drinks, coffee, breaded deep-fried chicken (or almost any other type of flavoured protein), are now appropriate at virtually any time, according to the whim of the individual, and in any combination. These he argues, constitute 'a nutritive medium within which social events occur', rather than the other way round. The meal, which once had a clear internal structure, dictated at least in part by the one-cook-to-one-family pattern, as well as by so-called 'tradition', can now mean different items, and a different sequence of items, for each individual consumer. The week's round of food, which once meant chicken (or some equivalent) on Sunday and fish on Friday, is no longer as stable, nor viewed as so necessary by the participants. At the same time, the year's round of food, which meant that certain vegetables and fruits, and the foods which are made with them, were only available (and only intentionally consumed) at certain seasons, certainly no longer obtains. Tradition still boosts the sale of turkeys at Christmas, and fish on Fridays perhaps, but certainly these 'substantializations' of the cycle of the year are no longer so strong.

While food consumption is much more individualized, and perhaps solitary, choices are still made from within a range which is predetermined by the limits of food technology on one side, and on the other by some notion of time constraint, or cyclicality, to some extent real, to a degree a remnant of tradition, and to an increasing degree, artifice, or contrivance. Mintz sees this as one of the fundamental reasons for the ubiquitous and various usage of sucrose in modern food, as well as for the very large increase in the consumption of sugars overall. The engineering of sensory experience which is involved in the creation of convenience foods finds sugar invaluable, not only for providing sweetness, but also for changing the texture of baked food (adding 'crispness'), adding body, or smoothing out a flavour, even if that flavour is not itself sweet (for instance, tomato sauce) (Fischler, 1983a, 1983b). The implication of Mintz's comments, although he does not draw it himself, is that 'convenience' in foods accomplishes an abstraction and compression of food qualities in a symbolic sense, as well as dispensing with the need for extended preparation. The use of sugar is one aspect of a general process whereby food and drink consumption is, increasingly, serving to make new sorts of social distinctions which are very different from those which we, in Western industrial societies, have held to be 'natural'. Relations between kin are no longer expressed in the same way within formal meals; notions of open and restricted spaces, and ideas of what behaviours are appropriate for what kinds of social spaces, have all come to be questioned, in addition to changed notions of the cycles of the day, the seasons, the year and the life span.

Mintz has raised the whole question of the significance of increased sugar use in the Western diet in a series of closely argued papers (Mintz, 1979, 1982). To begin with, he relates the increased availability of sugar and the increased sweetness of the Western diet to slavery, indentured labour and the production of primary commodities in the Third World. The symbolic meaning of sweetness in foods cannot take any priority over the social and economic systems within which these foods are produced and consumed, and any analysis of patterns of food consumption must take care to preserve an overview of all the dimensions involved, at both 'micro' and 'macro' levels.

DRINKING AND DEINDUSTRIALIZATION

Our own research on changes in patterns of drinking in North East England illustrates these processes very well. When we investigated changes in the market for ales in the area (Lesser *et al.*, 1983; Gofton, 1983) we found that this could be described in terms of two basic patterns. The 'traditional' pattern was to be found amongst older working-class men; this viewed ale drinking and the use of pubs as overwhelmingly a male pursuit; only working-class men learned to drink ale, they told us, and the drinking of ale was one of the ways in which such men distinguished themselves as a group. Women were excluded from the use of public houses except in restricted roles—only parts of the public house were open to them, and because of the women's presence these places were inappropriate as venues for 'normal male drinking'. Accredited status as a member of a male drinking group was only open to those who possessed the knowledge and skills acquired during socialization. These would include skills of discrimination, rules of comportment, reciprocity and appropriate forms of speech, behaviour and social solidarity.

It is clear that this pattern had developed historically as part of the culture of the working class in this region as a consequence of particular social, political and economic relationships arising from industrialization in the nineteenth century. Brian Harrison (1971) has argued convincingly that patterns of drinking in Victorian Britain can be seen as institutional responses to changing economic, social and political systems; pubs become, in this period, the essential and ubiquitous venue for working-class leisure, a way of escaping from the horrors of urban deprivation and the tyranny of the clock. This also becomes increasingly a male-dominated area of life as the family itself changes. Gender roles confine the wife/mother to the home; the state takes over the socialization of children in the educational system, and extends childhood (and training) beyond the norms found in preindustrialized societies. Ale itself comes to acquire symbolic associations with strength and virility; this symbolic meaning is accrued by combining folk beliefs about its nutritive efficacy, and thirst-quenching properties with long-established nationalistic associations, but also notions of the healthfulness of ale when compared with the dangers of other intoxicants. In his famous series of etchings, Hogarth contrasted the delinquency and depravity of Gin Lane with the rude wellbeing of Beer Street and its occupants (see Bailey, 1978; Smith, 1983).

The pub, and working-class ale drinking, develops during the nineteenth century as a result of political and economic decisions also. The Beer Act, 1832, was seen as a way of providing for the leisure of the new urban proletariat which avoided the dangers accompanying widespread use of stronger drinks, such as gin, but also contained and channelled the leisure activities of the new class, seen at the time as a potential source of social upheaval. At the same time, of course, new markets were created, and products based on the resources of the landowning aristocracy created immense fortunes for families such as the Guinnesses (Smith, 1979).

Our pattern of 'traditional drinking' can only be understood as the product of social and economic processes deriving from industrialization. It is itself a product of the new industries which grow up to cater for the needs of a mass, urban way of life, but in its practices it also embodies the social distinctions and relationships which are important to that social system, distinctions between the age and gender groups, and between the social classes which form the structure of this society. At the same time, working-class drinking as an institution cements the group together as its practices celebrate the values, image and identity of group members (Harrison, 1943).

We found, of course, that drinking patterns are now in the process of rapid change; but also that this change could only be understood in relation to the process of, as it were, 'deindustrialization'. For the young, drinking occupies a very different role in leisure—but leisure itself has a different significance, because work has changed. In the North East of England, the traditional heavy industrial workbase has been drastically affected by changes in the national and the world economy. Shipbuilding, coalmining and heavy industries such as steel and engineering have been decimated. Unemployment is chronically high, both amongst adults and the young. Apprenticeships in these industries, which used to form the main job opportunities for young men, have been severely cut, and what jobs there are for young people tend to be outside of these traditional industries.

The implications of these changes are wide ranging; so far as leisure drinking is concerned, young men no longer enter the world of work, and consequently membership of adult male drinking groups. They no longer learn to drink in the company of older adult males to the same degree that they did. At the same time, the population of towns and cities, which used to be concentrated in the centres, has been moved out to housing estates on the outskirts. Pubs and clubs, which were sited in the centres to cater for the adult male workforce, have now been forced to cater for a new market. During the day, they sell meals and a range of drinks to increasing numbers of (largely middle-class) business and commercial people who work in the town centres, while at weekends and at night, they are transformed into 'young' pubs catering, with bright lights and swish décor, for those who have the time, mobility and inclination to travel in for their entertainment.

These are very different from the traditional pubs; different drinks are sold, and different ends are pursued by the customers. For the young, drinking is only a part of public house usage; the main reason is to see other people—new faces rather than simply close friends—and also to be seen. Drinking and pub use are, for the young, very much a matter of fashion; traditional ales are largely rejected because they convey a class-based parochialism which has no place in their self-image. The new drinks, like ale in the old pattern, are props in the staging of a self-image (Goffman, 1969). Whereas the image of the traditional drinker was founded largely on social distinctions based on work, however, for the 'new' drinkers, styles of leisure are a far more important factor.

Leisure consumption is far more varied for the young, with clothes, fashion, sport and music as just some of the important areas providing cultural resources which are employed in the constitution of the self. The essential feature of traditional drinking was the way in which it embodied practices which reinforced the solidarity of the (work) group; in contrast, new drinking is individualistic, and a mass rather than a small-group phenomenon. The pub, as such, is no longer the focus of drinking by groups, but simply a staging point for a transient population; town centres themselves are the venues for displays of conspicuous consumption, with pubs forming 'rounds' which are visited in the course of an evening. Drinking groups are composed of both males and females, without the same hard distinctions as to what should be appropriate drinks for one or the other—women still do not, by and large, drink ales, but lager, which used to be seen as an inappropriate drink for men,is now consumed by younger male drinkers. Twenty years ago it was 1 or 2 per cent of the market for drink in the UK; that figure is now approximately 40 per cent. Drinks such as cider, which were previously a minority taste for the same reason, have similarly been taken up by young people of both sexes.

These changes must be seen as indivisible from the decline of traditional forms of work, and the consequent breakdown of relationships between the generations based on work. There are other changes, of course. The older generation has been affected by the breaking up of old communities and the movement of population away from the city centres, by the changing position of women in society and the changing nature of family relationships and conjugal roles, and by the 'privatization' of the working-class family which has led to the growing importance of the home as a venue for leisure. The fundamental changes in drinking, however, can only make sense as a consequence of adjustment to the decline of traditional forms of work, and the growing significance of leisure consumption as a medium for the expression of individual and group identity. This involves, to a far greater degree, the appropriation of social spaces for commercialized mass consumption of goods and services, and the creation of new markets which cater for changing, fashion-based tastes which are interrelated elements of the new leisure-dominated lifestyles of the urban population.

This article has argued for a social analysis of food consumption which attends to the relationship between food consumption and the nature of the groups within which it occurs. Food consumption can only be understood as a cultural institution which reflects the main features of that group life—it expresses relationships between members of those groups, and also between the groups of which society is composed. But, of course, it also relates to other social institutions; the production of food, how it is distributed and divided, the technology which is employed in food manufacture, and the economic and political relations which are involved must also form central elements in any analysis of food consumption. The issues involved constitute a formidable agenda. Research into the changing forms of food consumption, the changing tastes of food consumers, and the problems of forming appropriate food policies and production and marketing strategies can ill afford to ignore the social dimension.

REFERENCES

Allen, D.E. (1968) *British Tastes: an Enquiry into the Likes and Dislikes of the Regional Consumer*. Hutchinson, London.

Angel, L.J., and Hurdle, G.E. (1978) 'The nation's food: forty years of change', *Economic Trends* (294).

Arens, W. (1979) *The Man-Eating Myth—Anthropology and Anthropophagy*. Oxford University Press, New York.
Avery, G. (1984) 'Children's corner', *New Society*, December, **170** (1148). Special Food Supplement.
Bailey, P. (1978) *Leisure and Class in Victorian England.* Routledge & Kegan Paul, London.
Barthes, R. (1979) 'Towards a psychosociology of contemporary food consumption'. In *Food and Drink in History: Selections from the Annales Economics, Societies, Civilisations,* ed. R. Forster and O. Rarum. Johns Hopkins University Press, Baltimore.
Beiger, J. (1970) *Ways of Seeing*. Penguin Books, Harmondsworth.
Bettelheim, B. (1970) *The Informed Heart*. Paladin, London.
Bourdieu, P. (1984) *Distinction: A Social Critique of the Judgement of Taste*. Routledge & Kegan Paul, London.
Braudel, F. (1973) *Capitalism and Material Life, 1400-1800*. Weidenfeld & Nicolson, London.
Brecken, W. (1975) *You Are What You Eat*. Thames and Hudson, London.
Burnett, J. (1979) *Plenty and Want: A Social History of Diet in England from 1815 to the present day*. Scholar Press, London. (Revised edition).
Chaney, D. (1983) 'The department store as a cultural form', *Theory, Culture and Society,* **1** (3), 22-31.
Chisnall, P.M. (1975) *Marketing: A Behavioural Analysis.* McGraw Hill, London.
Detienne, M. (1972) *Les Jardins d'Adonis*. Paris.
Detienne, M. (1977) 'La Viande et le sacrifice en Grece ancienne', *La Recherche*, February (75), 152-160.
Douglas, M. (1966) *Purity and Danger*. Pelican, London.
Douglas, M. (1976) 'Culture and food'. In "Culture; Essay on the Culture Programme". Russell Sage Foundation Annual Report 1976-77 pp. 55-81. Reprinted in Freilich, M. (ed.) (1983) The Pleasures of Anthropology, Mentor Books, New York.
Douglas, M. (Ed.) (1984) *Food in the Social Order: Studies of Food and Festivities in Three American Communities.* Russell Sage Foundation, New York.
Douglas, M., and Isherwood, B. (1979) *The World of Goods.* Allen Lane, London.
Dumont, L. (1970) Homo Hierarchicus: *The Caste System and its Implications*. Weidenfeld & Nicolson, London.
Elias, N. (1978) *The Civilising Process* (2 vols). Vol I *The History of Manners*. Vol II *State Formation and Civilization.* Basil Blackwell, Oxford.
Elias, N. (1983) *The Court Society*. Basil Blackwell, Oxford.
Erasmus, D. (1530) *De Civilis Morum Pueritium*, cited in Elias (1978).
Evans-Pritchard, E. (1940) *The Nuer: A Description of the Modes of Livelihood and Political Institutions of a Nilotic People*. Clarendon Press, Oxford.
Fischler, C. (1983a) 'L'Empire de la douceur'. *Le Group Familial*, Avril, **99**.
Fischler, C. (1983b) 'Le Ketchup et la Pillule', *Perpective et Sant,́* **25**, 110-119.
Fischler, C. (1984) 'Fatness, gluttony and sharing: on the social meaning and management of male obesity and fatness'. Paper presented at Symposium on Food Sharing, held at Werner-Reimers Stiftung, Bad Homburg, Germany. December.
Flandrin, J.L. (1979) 'Families in former times'. Kinship, *Household and Sexuality*. Cambridge University Press, New York.
Foucault, M. (1967) *Madness and Civilisation: A History of Insanity in the Age of Reason*. Tavistock, London.
Foucault, M. (1970) *The Order of Things: An Archaeology of the Human Sciences*. Tavistock, London.

Furnivall, F.J. (1868) *Early English Meals and Manners.* Early English Text Society, London. No. 32.
Goffman, E. (1968) *Asylums.* Penguin, Harmondsworth.
Goffman, E. (1969) *The Presentation of Self in Everyday Life.* Penguin, Harmondsworth.
Goffman, E. (1972) *Relations in Public Micro Studies of the Public Order.* Penguin, Harmondsworth.
Gofton, L.R (1983) 'Real men, real ale', *New Society,* **66** (1096), 2713.
Goody, J. (1982) *Cooking Cuisine and Class. A Study in Comparative Sociology.* Cambridge University Press, New York.
Hardy, T. (1874) *Far from the Madding Crowd.*
Harris, M. (1975) *Cannibals and Kings.* Random House, New York.
Harrison, B. (1971) *Drink and the Victorians.* Faber and Faber, London.
Harrison, T. (1943) *The Pub and the People: A Worktown Study By Mass Observation.* Redwood, London.
Henry, Jules (1966) *Culture Against Man.* Pelican, London.
Hopkins, H. (1985) *The Long Affray: The Poaching Wars in Britain, 1790-1914.* Secker & Warburg, London.
Kornhauser, A., and Lazarsfeld, P. (1935) 'The analysis of consumer actions'. Reprinted in *The Language of Social Research,* ed. Lazarsfeld and Rosenberg (1955). Collier-Macmillan Press, Toronto.
Lesser, D., Ritson, C., Ness, M., and Gofton, L. (1983) 'The North East beer drinker'. Unpublished report prepared for Scottish and Newcastle Breweries Ltd.
Lévi-Strauss, C. (1969) *The Raw and the Cooked.* Harper and Row, New York.
Lévi-Strauss, C. (1978) *The Origin of Table Manners.* Jonathan Cape, London.
Mintz, Sidney (1979) 'Time, sugar and sweetness', *Marxist Perspectives,* **2**, 56-73.
Mintz, Sidney (1982) 'Choice and occasion—sweet moments'. In *The Psychobiology of Human Food Selection*, ed. L.M. Baker. AVI, Westport, Conn.
Murcott, Anne (ed.) (1983) *The Sociology of Food and Eating. Gower*, Aldershot.
Nicod, M. (1979) 'Gastronomically speaking'. In *Nutrition and Lifestyle*, ed. M.R. Turner. Applied Science Publishers, London.
Oddy, D., and Miller, D. (eds.) (1976) *The Making of the Modern British Diet.* Croom Helm, London.
Palmer, A. (1952) *Movable Feasts: Changes in English Eating Habits.* Oxford University Press, Oxford.
Reissman, D., Glazer, N., and Denney, R. (1950) *The Lonely Crowd.* Yale University Press, Newhaven.
Reynolds, F.D., and Wells, W.D. (1977) *Consumer Behaviour.* McGraw-Hill, New York.
Sen, Amartya (1981a) *Poverty and Famine: An Essay on Entitlement and Deprivation.* Oxford University Press, New York.
Sen, Amartya (1981b) 'Ingredients of Famine Analysis: Availability and Entitlements', *Quarterly Journal of Economics*, August, **XCVI** (3), 433-464.
Smith, M. (1979) *Brewing Industry Policy: The Public House and Alcohol Consumption Patterns in the UK.* Centre for Leisure Studies, Salford University.
Smith, M. (1983) 'Social usage of the public drinking house: Changing aspects of class and leisure', *British Journal of Sociology,* **XXXIV** (3), 367-385.
Svenbro, J. (1984) 'The division of meat in classical antiquity'. Paper presented at Symposium on Food Sharing held at Werners-Reimers-Stiftung, Bad Homburg, Germany, December.
Tannahill, R. (1973) *Food in History.* Stein and Day, New York.
Trevelyan, G.W (1979) *English Social History.*
Turner, Barry (1982) 'The discourse of diet', *Theory, Culture and Society,* **1** (1), 23-32.

Turner, Barry (1984) *The Body in Society*. Basil Blackwell, London.
Zaltman, G., and Wallendorf, M. (1979) *Consumer Behaviour: Basic Findings and Management Implications*. Wiley, New York.

The Food Consumer
Edited by C. Ritson, L. Gofton and J. McKenzie

CHAPTER 8

An Integrated Approach—With Special Reference to the Study of Changing Food Habits in the United Kingdom

JOHN MCKENZIE

This book is about what makes the food consumer tick! It considers how individuals determine their food consumption patterns, what influences their choice and, perhaps most important of all, how we can predict changes in these patterns that will occur over time.

Part II studies people's attitudes and behaviour relating to food, and the preceding chapters have looked in some depth at the influences of economic, psychological, anthropological and sociological determinants upon choice.

But why is such an analysis of consequence? In fact, it goes well beyond the fairly obvious issue of accurately predicting the right products for the food manufacturer to produce, although it must be stressed that the high failure rate of new food products and the financial implications resulting therefrom do, in their own right, merit further study. Furthermore, in a world where so many have so little, be they the unemployed in Britain or the starving in Ethiopia, provision of food for waste is an unacceptable tenet!

In the context of Western society, however, there is a further and perhaps even more important concept. As John Yudkin, Professor of Nutrition and Dietetics at the University of London, said in 1964, 'The experience of what determines food habits and how they can be changed is of concern to the health of all mankind.' (Yudkin, 1964)

This statement must be seen in the context of a review of the development of nutritional science (in reality a twentieth-century discipline), and its success, or lack of it, in directing food consumption patterns to bring optimum health and alleviate disease. The evidence is not encouraging. While certain parts of the world have moved away from the diseases associated with nutritional deficiency (albeit probably as a result of affluence rather than of consumption patterns directed specifically to meet our nutritional needs), these problems have been replaced by a series of food/nutrient excesses which appear to be at the very least codeterminants of many key health problems such as heart disease. And there is little evidence to suggest that changing food habits to resolve these relatively new health problems is easy to achieve.

Let us examine this matter in a little more detail. In the context of most of Europe, the development of nutritional science may be reflected by three periods of significant contrast, each identifying our inability to influence choice to any degree.

During the first period (let us say from the beginning of the twentieth century up to the beginning of the Second World War) scientists and clinicians in general tended to believe that if they simply set out to pursue research and publish their findings, general practitioners and others dealing with the general public would use the knowledge to cure, and possibly prevent, ill-health. However, by the beginning of the Second World War, it had become apparent that this did not really work to any great extent—except possibly when clinicians were aware of, and able to deal with, some specific deficiency conditions.

The second period—coinciding with the years of the Second World War—was one of strict controls whereby as a result of rationing, scarcities and a whole range of propaganda activities, the government of the UK attempted to control food habits in the interests both of maintaining an acceptable food supply and of ensuring health. Although the distribution of food within the family presented some problems (with father often getting the scarce protein items intended for pregnant and lactating mothers and young children), to a large extent this system worked. However, it was clearly not possible to pursue such a procedure for any period following the war (as the postwar Labour Government in the UK found to its cost).

In any event, a third new situation was rapidly emerging. On the one hand, growing prosperity was expected to resolve most nutritional problems which had been thought to result from serious poverty. On the other hand, it was increasingly believed that if people (now better educated) were told what they should eat to obtain a nutritionally more desirable diet, they would behave appropriately. Thus we entered the 'education and open purse' period.

But alas, it soon became apparent that this did not deal adequately with the situation. Prosperity certainly brought with it changing food patterns, but these patterns were not necessarily nutritionally desirable. Indeed, the nutrition of an affluent society presented all sorts of new problems related to eating too much (either in total or of certain nutrients). And people did not seem to modify their behaviour in the light of increased nutritional knowledge.

Thus, in 1983 the National Advisory Committee Report on Nutrition Education set the following targets for the UK:

10 per cent reduction in total fat intake

15 per cent reduction in saturated fatty acids

10 per cent reduction in sugar

More exercise

25 per cent more dietary fibre

10 per cent reduction in salt.

But the stark reality is that whilst we are increasingly coming to understand desirable nutritional objectives, our knowledge of how to influence the consumer to seek to attain these objectives is not much more advanced than it was in the 1930s.

Thus, just as Margaret Mead in the 1940s could talk about the need for fundamental study of the factors influencing food choice, so the claim could be repeated with legitimacy by Burgess and Dean in the 1960s and Yudkin and McKenzie in the following 20 years. (Yudkin and McKenzie, 1964; Yudkin, 1978; McKenzie, 1974, 1981). Whilst the problem now is readily acknowledged by nutritionists and social scientists alike, Roseamann found equal justification in 1985 for making a similar lament!

None of this is to suggest that food habits are not changing. Personal experience in the UK would identify easily, for example, the growing markets for fruit juice, yoghurt, chilled desserts and poultry foods, and the decline in consumption of, say, fish and tinned foods. But what evidence there is suggests that these changes have come about primarily because the consumer wishes for change rather than because the nutritionist (or even to some extent the food manufacturer) has sought it. Perhaps people today are more health conscious but the movement of attitude to behaviour change is unbelievably slow.

In part at least the difficulty in the professional attempting to create change in consumption patterns for health reasons, lies in the complexity of the study necessary to understand the influences at work in food choice and in the fact that such study embraces a wide range of science and social science disciplines. Few people have felt able to embrace the totality of the issue.

The chapters in Part II of this book dissect key social science concepts that need to be reviewed to aid our understanding of what determines food choice. Here I seek to place these various elements in an interrelated context, and to add other relevant dimensions of consequence.

First, as all would admit, the relative price of foods acts at least as a crude determinant of choice. Elsewhere (McKenzie, 1980) I have suggested that this can best be described by the analogy of placing all available foods on a table. The very price of certain of those foods (be it caviar, fresh strawberries, fillet steak or even milk) may for different groups lead to them being removed from the table. But the selection from what remains is probably based less upon individual comparative price than by relative preference, be it fundamentally determined by taste, food imagery, psychological need or whatever.

However, such a view needs qualifying. Complementary products and directly competitive products obviously interrelate. The price of butter has an effect on demand for margarine and bread upon the demand for jam. A separate point which reflects sadly on the 1980s is that recession has had a significant effect on the foods that remain 'on the table'. This is not necessarily because of their cost in absolute terms. The fact that, as society becomes richer, the percentage of money spent on food declines is well established—the issue is discussed in some detail in Tangermann's chapter. What is less well recognized is that the food budget is one of the few possible sources of reduction in family expenditure in times of short-term hardship in modern society.

The reason may be explained at three levels. As costs generally rise the household is faced with fixed parts of the budget that cannot be reduced—it is difficult to avoid continuing to pay the mortgage bill, even the rates if they have gone up, the electricity or the HP on the car. Thus, if real income declines, then the only short-term opportunities for rebudgeting lie within those fields where there is some flexibility—luxury items, entertainment, holidays, savings and food. There is also clear evidence that during such periods of inflation coupled with recession, most housewives do not obtain *pro rata* housekeeping budget increases.

At a second level, since the food budget provides room for identifying flexibility (in terms of the food left on the table), it is possible to move to cheaper brands, own-label products, even close substitutes, without apparent hardship. (See Figure 8.1.)

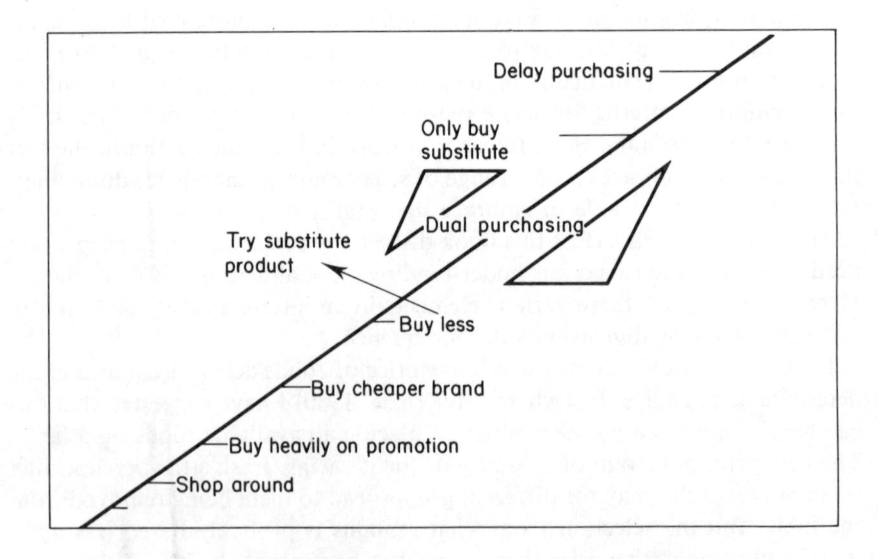

Figure 8.1 Modification of purchasing behaviour (after Ramsbottom).

Lastly, there may well be room to eliminate wastage. A tendency to buy too much appears to be the only explanation for the apparent significant jump in food consumption quantities during the whole of the 1970s. Convenience foods do not have so much wastage built in! And refrigerators and freezers make storing considerably easier.

However, this surprising national (and, I suspect, international) wastage factor masks one even more serious concern. In no other field does the recent recession more seriously divide the 'haves' from the 'have nots' than in the field of food. In 1980/1981 it was graphically shown that during a 12-month period there was an increase of 9 per cent in money spent on grocery items in areas of low unemployment compared to 1 per cent in areas of high unemployment. Behind this expenditure lie significant variations in demand for different sorts of product between the areas of high and low unemployment.

A second element I should like to mention in our interrelated context is that anthropological and sociological studies highlighted in the chapters by Murcott and Gofton have shown fundamental influences of religion and food imagery upon choice. The former, involving the rejection of particular foods (e.g. pork) in certain countries or amongst certain races, is well documented, but the latter issue is perhaps less well known. Yet anthropologists have shown, on an international basis, the widespread folklore and traditions associated with food. Thus, in some cultures in developing countries a relatively abundant item of food may be unacceptable, either absolutely or at certain times such as during pregnancy or lactation, when the patient is ill, or during times of mourning.

In parts of Latin America foods are divided into 'hot' and 'cold' items. The former may be consumed at any time; the latter only by healthy adults, not in the morning and only if their effects are neutralized by hot tea.

In China food is divided traditionally into two broad categories: fan (grain and starch foods) and tsai (vegetables and meats). Fan is regarded as fundamental, indispensable food whilst tsai are the secondary, luxury foods. They are cooked separately and must be served in separate utensils. The infinite permutations of food, within and between categories, are seen not only to affect the individual's health but his very mood and personality.

In the UK many traditional attitudes have been passed down the centuries. The earliest book in English on diet, published in 1633 (Hart, 1633), identifies many of the characteristics of food imagery found in current times in many parts of the developing world. It is also possible to see elements of these images continued in current consumer attitudes and thought in the UK. New food imagery has also emerged. Thus, people talk about 'natural foods', 'roughage', 'slimming foods', 'energy foods' and so on. Such imagery may inhibit change if a new recommendation appears to conflict with a traditional view of particular foods. Or it may encourage change if a new move is supported vigorously by the community as a whole or by a specialist group.

Equally, certain forms of behaviour become fashionable. By adhering to them, individuals associate with people of like mind and may be seen as leaders of fashion or members of a particular group or class. An example would be the growing interest in foreign dishes encouraged by travel abroad.

Third, and in a comparable context, psychological influences not surprisingly come to bear. Behind the detailed analysis identified in previous chapters, I believe at least five fundamental laws may be identified:

(1) Food preparation, storage and consumption act as an aid to feelings of security.
(2) Food selection and preparation act as a direct substitute for maternal creativity.
(3) Food choice demonstrates group acceptance, conformity, prestige.
(4) Food compensates for denial so is used as a support during times of crisis.
(5) Our choice of food and drink acts as a means of demonstrating mood and personality.

Often many such issues produce a wry smile or a cynical look and are seen as 'quack medicine' or 'Readers' Digest psychology'. But I can think of no simpler way of reinforcing the cultural, historical and psychological significance of food than by an experiment I conduct each year in the lecture theatre. I indicate that I will bring in two unusual foods for my students to try, and they are asked to select which they would prefer to taste. The first is a new variant of a loaf based on leaf protein—its only unusual characteristic is that the loaf is green! The second is a Philippine delicacy recently imported—the live foetuses of mice dipped in honey. Which will the students choose to try? Not surprisingly, 99 per cent go for the green loaf (and one wonders a little about the psychological state of health of the 1 per cent who go for the foetuses dipped in honey).

The students have seen neither food and both descriptions are factual and unemotive. The reality is that there are enormous psychological and cultural influences built up over thousands of years which make us overtly much more suspicious about unusual animal than vegetable products. Anthropologists and psychologists could describe the reason for this at length. I make the point here simply to demonstrate the potency and cogency of inherent and acquired prejudices (or preferences).

So far in this chapter my aim has been to put into perspective and interrelate some of the fundamental disciplines which influence choice, and which have been covered in more depth in the previous four chapters. The main purpose of the remainder of this chapter is to indicate that there are a number of other factors no less profound which have perhaps an even more fundamental influence upon choice.

Perhaps the most obvious is the whole issue of availability, not determined by price, but by accessibility. The possibility of purchasing a particular brand of a product depends upon its availability in the shops. And there is increasing evidence that, with the reduction of the number of retail outlets and the concentration of the great majority of items in the hands of a limited number of retail companies, the real availability depends upon the policies pursued by the main buyers for these stores. In reality, choice may be in the hands of perhaps fewer than one hundred people.

This has particular influence where a new product is concerned, or where, say, a third company seeks to enter a market dominated by two existing brand groupings in a proposed field and which takes considerable shelf space. The classic example has been the inability of any other manufacturers successfully to break the dominance of two major manufacturers in the baby foods market where, if you are to place one product on a shelf, you have to provide a range that takes considerable space. It is, perhaps, fortunate that overall this limited number of buyers are often in very significant competition with each other and rarely act in unison.

Another fundamental influence exists at a physiological level within ourselves. In our relatively prosperous Western society, consumption of food for most of us has now reached a high plateau: a plateau above which it is physiologically almost impossible to go. Therefore, as we get richer we do not increase consumption of all foods in total, but begin to change the ratio of the quantities of the various foods we eat. Thus, the richer groups of the population consume more of some foods, such as meat and fruit, but in consequence less of others, such as bread and potatoes.

Because of this new behaviour trend, all foods now compete with each other. Increased consumption of one can only be at the expense of another. If we are to understand the dynamics of both present and future trends in consumption, it is necessary to understand the basis of the interrelationship between various foods. The first step in such a study is to build up a model of the various food categories and their interrelationship. We must think of food categories not so much in terms of what might be a technically appropriate breakdown in the eyes of the food manufacturer or nutritionist as in terms of the public's conception of them. By this I mean that once the housewife has decided on the structure of a meal in terms, for example, of 'meat' and 'vegetables', one needs to know what she regards as possible alternative foods within these categories.

In Figure 8.2 I have attempted a breakdown of foods into these various categories. The figure represents the present position, but with the exception of the occasional introduction of new foods and changes in the relative consumption of the different foods within a category, there is little to suggest that there has been much change within the categories over the years. The list for each category is not meant to be comprehensive but merely to indicate some of the major alternatives within each group. Some categories are clearly interrelated. Thus, bread and potatoes might be regarded by some as in the same category, or the division between some fruits and vegetables held to be marginal. In such cases I have placed arrows on the figure indicating a possible overlap.

Figure 8.2 Public conception of food categories.

1 Meats	2 Bread	3 Potatoes	4 Vegetables	5 Fruit	6 Puddings	7 Cakes
Meat	Bread	Potatoes	Peas	Apples	Tinned fruit	Cakes
Fish		Rice	Cabbage	Bananas	Fruit puddings	Biscuits
Eggs		Spaghetti	Tomatoes	Oranges	Rice	
Cheese		Pasta	Beans	Pears	Jelly	
Poultry			Onions			

8 Confectionary	9 Sugar	10 Beverages	11 Milk	12 Spreads	13 Breakfast Cereals
Sweets	Sugar	Tea	Milk	Butter	Cornflakes
Chocolates	Sugar subs.	Coffee		Margarine	Porridge
		Coca		Jam	
				Marmalade	
				Pastes	

Such a figure may be helpful at several levels when studying trends. In the first place it will indicate the intergroup relationship of the various foods. For example, it might show that the demand for 'spreads' has not changed, but that the consumption of butter has risen directly at the expense of margarine. Second, it may help to establish the intergroup relationship of foods both on a complementary and inverse basis. At a complementary level a decrease in the consumption of stewed fruit may directly lead to a decline in the consumption of custard. At an inverse level there may be a relationship between the increase in the consumption of meat and the decline in the consumption of bread. I believe that we shall come to view current trends in food consumption much more realistically if we study the whole picture in this way.

A third basic factor influencing selection relates to recent demographic changes. In Figure 8.3, the changes in the lifespan of women during the last 100 years and in the relative periods of their life when they are concentrating upon child bearing and rearing are identified. It will be evident that 100 years ago, by the time the children had 'left the nest', the woman's life was virtually over. The harsh reality was that on average the mother's death virtually coincided with her youngest child reaching 10 years of age. As a result of dramatic improvements in the environment and in our knowledge and ability to deal with disease, the lifespan of the average woman has nearly doubled during the last 100 years and she can now expect to live 40 years after her youngest child reaches the age of 10.

The ramifications of this change, especially when related to the increasing knowledge and efficacy of contraceptive methods, have been enormous. And it is in this context that women's changing attitudes towards opportunities for

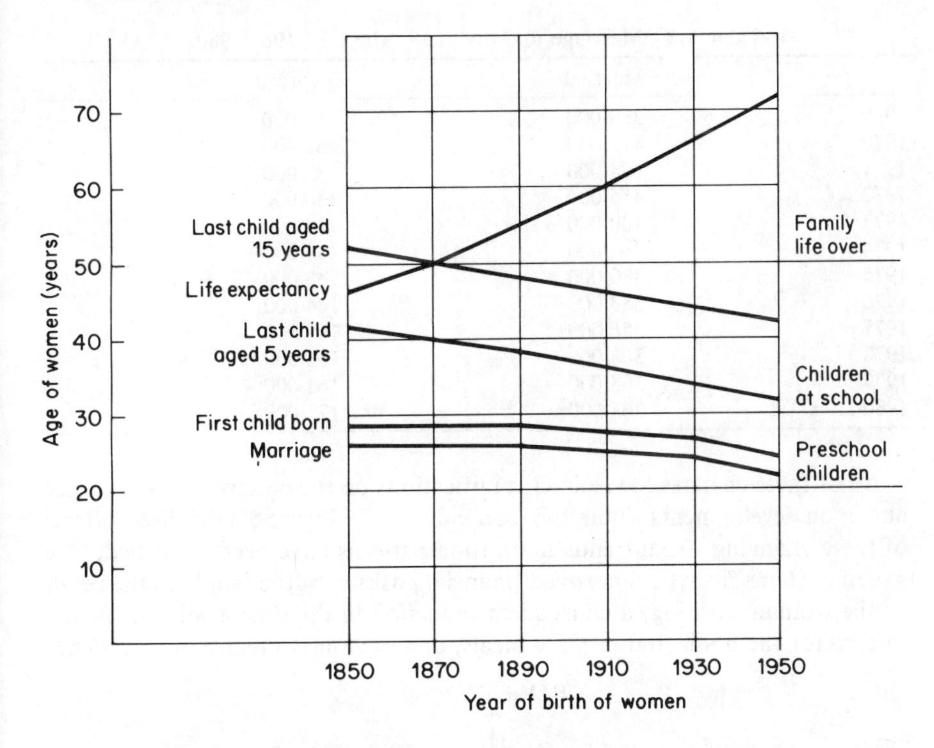

Figure 8.3 Changing position of women resluting from the decrease in the size of the family, age of women at childbirth and increase in life expectancy of infants.

work have developed. In the same way it probably explains the present day breakdown in the stability of marriage. There is no longer the 'cement' of young children to keep the partners together. The economic independence the woman derives for herself gives her sufficient security to step aside from either a contracted or a common law marriage if she believes it has not been or is no longer satisfactory.

In Tables 8.1 and 8.2, the exact nature of these steps is more specifically identified. Thus, one can see a significant change in the size of the household over the last 20 years and that one in three marriages is now likely to break up.

Table 8.1 Size of UK households, 1961, 1971 and 1981.

	1961 %	1971 %	1981 %
One person	12	18	23
Two persons	30	32	32
Three persons	23	19	17
Four persons or more	35	31	28
Average household size (number of people)	3.1	2.9	2.7

Table 8.2 Marriage and divorce in the UK, 1969-1980.

	Married	Divorced
1969	369 000	53 000
1970	415 000	62 000
1971	404 000	83 000
1972	415 000	130 000
1973	400 000	128 000
1974	384 000	134 000
1975	380 000	140 000
1976	358 000	144 000
1977	356 000	165 000
1978	368 000	163 000
1979	368 000	163 000
1980	363 000	177 000

All of these changes have direct ramifications on the pattern of food choice and upon developments within the food industry. In Figure 8.4 the direct effects of these changing social trends upon food patterns have been identified. The situation stems from the improved financial position of the family as the result of the woman working, a consequent reduction in the time available for her to care for the home and prepare meals, and obviously greater independence

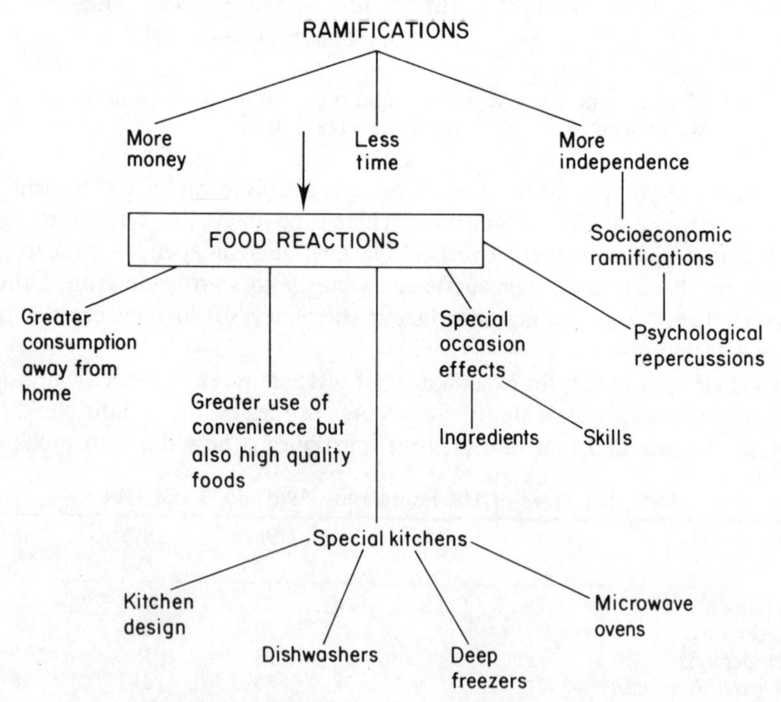

Figure 8.4 Ramifications of women at work.

of all sorts directly derived from her work activity. The effects upon food choice may be summarized as follows:

(1) A significant growth in the quantity of food consumed away from the home in restaurants, snack bars, etc.
(2) The greater use of convenience foods/ready-prepared meals for everyday occasions.
(3) The consumption of these everyday meals in very casual circumstances (see Figure 8.5).
(4) A growing distinction between foods for nourishment and foods for fun (see Figure 8.6).

Figure 8.5 Distinction between formal and casual meals.

Formal meals	*Casual meals*
Very occasional, e.g. if Sunday lunch or visitors	Very frequent meals
All family eating together	Often people eating alone
More formal surroundings	In kitchen/lounge
Expensive/'special' meals	Ordinary meals/snacks
Lot of preparation by housewife	Little preparation by housewife/ significant convenience role
Knife and fork	Straight from container

Figure 8.6 Distinction between foods for nourishment and foods for fun.

Foods for nourishment	*Foods for fun*
Specific meals and parts of meals, e.g. breakfast, main course at dinner	Remainder of meal/snack occasions
Designated nutrient intake	No concern about nutrient value (probably better if no nutritional value)
Concern about consumption	No concern about consumption
Need for reassurance from manufacturer	Other than not harmful, no questioning of manufacturer
Intrinsic merits in kudos of product, but should also taste OK	Fun to look at, nice taste, good image

Equally, the growth in prosperity, the changing size of the family and the increasing proportion of women at work have led to dramatic changes in the food industry over the last 20 years. These changes have themselves facilitated even further growth in certain trends. Some of the more obvious manifestations within the food industry have been:

(1) Changing methods of preservation (and in particular the emergence of freezing).
(2) Rapid growth in convenience products and ready-cooked products and meals which simply need reheating.
(3) The growth in ready-to-eat products (a classic example being the items now available from the chilled cabinet, e.g. yoghurt, cream desserts, etc.).
(4) The enormous extension of the snack products market.
(5) Provision of appropriate quantities for the small family.
(6) The provision of appropriate psychological support.

Thus, at a price (which to a two-wage-earner household is acceptable) the food industry has taken over many functions previously handled by the housewife, has facilitated her demand to work and has tried to satisfy some of her latent psychological needs. While in some areas the industry may be reasonably criticized, this in my view is not one of them. If housewives buy tinned custard and virtually transfer menu planning to the food industry, it is primarily because of their fundamental desire to transform the nature of their lives. The implications upon food choice are of a coincidental, secondary nature, rather than a primary influence.

There remains one further strong trend with a marked influence upon choice: there is increasing evidence to suggest that very recently health perceptions have come to play an increasing role in choice in Britain, Western Europe and North America. This is not so much based on technical nutritional knowledge (although Thomas, 1980, and others more recently have indicated rapidly growing detailed knowledge) as on general concern about health and heart disease, high blood pressure, cancer (especially the role of additives, etc.).

I believe it is possible to see this influence manifested in the rapid growth of perceived 'natural' products (such as yoghurt, fruit juice, bran products) and the strong movement away from sugar and animal fat. Is this trend a new, passing fancy? I suspect not—here I believe it reflects a new awakening to the issue of the influence of food on health. And it is being reinforced by the government and manufacturers by labelling regulations (aimed at giving more specific content information), health food 'bars' or sections of shops, and appropriate products and advertising.

But we should not get this changing picture out of perspective! It remains self-evident both from an analysis of consumption patterns and the admissions of consumers themselves that health matters still do not top the list of priorities determining choice (see the reasons given for selecting the foods we eat, given in Table 8.3). There is also the magic use of the 'compensation theory'—that is, four heavy meals can be absolved by one salad lunch, or heavy drinking sessions at the weekend compensated for by mid-week abstinence. It is a sad but expected phenomenon of life that a large amount of excess is expected to be absolved by a small amount of penance—a factor no less true in nutritional than in religious circles!

Table 8.3 How important are each of the following when selecting the food you eat?

	Very important	Quite important	Not very important	Not at all important	Don't know
Taste	83%	14%	1%	0%	0%
Value for money	67%	26%	5%	1%	1%
Nutritional value	62%	27%	9%	2%	1%
Price	55%	33%	10%	2%	1%
Family influence/ preference	52%	30%	10%	5%	3%
Convenience	24%	41%	29%	5%	0%

Source: Gallup.

REFERENCES

Burgess, A. and Dean, R. F. A. (1962) (eds.) *Malnutrition and Food Habits*, Tavistock Press, London.

Hart, J. (1963) *Kainikh on the Diet of the Diseased.* Robert Allot, London.

McKenzie, J. (1980) Economic influences on food choice, in Turner, M. (ed.) *Nutrition and Lifestyles.* Applied Sciences Publications Limited, London.

McKenzie, J. (1981) What is Social Nutrition? *Nutrition Bulletin,* British Nutrition Foundation, London.

Mead, M. (1943) *The Problem of Changing Food Habits.* National Academy of Sciences, Washington DC.

National Advisory Committee on Nutrition Education (1983) *Proposals for Nutritional Guidelines for Health Education.* Health Education Council, London.

Roseamann (1985) *New Society.*

Thomas, J. E. (1980) Relationships between knowledge about food and food choice, in Turner, M. (ed.) *Nutrition and Lifestyles.* Applied Sciences Publications Limited, London.

Yudkin, J. (1964) The need for change, in Yudkin, J and McKenzie, J. (eds.) *Changing Food Habits,* MacGibbon and Kee, London.

Yudkin, J. (ed.) (1978) *Diet of Man: Needs and Wants,* Applied Sciences Publications Limited, London.

Yudkin, J. and McKenzie, J. (1964) Conspectus, in Yudkin, J. and McKenzie, J. (eds.), *Changing Food Habits,* MacGibbon and Kee, London.

PART III
Presenting Food

The Food Consumer
Edited by C. Ritson, L. Gofton and J. McKenzie

CHAPTER 9

Researching The Food Consumer

TECHNIQUES AND PRACTICE IN THE UK AND NORTH AMERICA

DAVID LESSER, DAVID HUGHES AND DAVID MARSHALL

INTRODUCTION

Man, or more accurately, woman is probably more closely observed and persistently measured in the role of a food consumer than in any other aspect of daily life. There is always someone somewhere who wants to know what she eats, when, where and in what combinations, where and when she buys the food, and, especially, why she chooses one foodstuff or one brand rather than another. The sponsors of these inquiries are mainly either governments and other relatively disinterested public bodies or firms and organizations with a clearly acknowledged profit motivation.

The government-sponsored studies are mostly those which aim at a comprehensive description of what is eaten, by whom and in what circumstances, and in many cases also of what is bought and how much is paid for each item. In principle, the findings are designed to be used as a basis for the formation of policy on nutrition, food manufacturing and distribution or agricultural production, and the price and quality information help to make up indices of the cost of living and thus could contribute to the formation of wider economic policy. Commonly, though, the findings are simply published as statistics, which can be used to support as many opposed arguments as most other public statistics, so that the findings do little more than offer the security of knowledge, without any policy changes being made as a consequence of the knowledge. Even if no action follows directly from the research, however, it may be assumed that the findings do show whether any action is needed; they are in reality, therefore, guides to action. Examples are the Ministry of Agriculture, Fisheries and Food's National Food Survey in the UK and the US Department of Agriculture sponsored National Menu Survey in the USA, both of which are discussed in detail later.

The commercial research, too, is largely concerned with describing food consumption and purchase behaviour, but usually more selectively and in more detail than is offered by the government publications; it is also mainly concerned with explaining that behaviour. The findings of commercial market research are used in attempts to reinforce or change consumer behaviour, perhaps to induce the consumer to buy more of or pay more for one product at the expense of a competing offer. What is needed is precise, up-to-the-minute information on what brands the consumer is choosing and why she is choosing those brands. The findings are nearly always put to use; for why else would the expense be undertaken? This research ranges from gathering usage or attitude data from syndicated consumer research panels (such as the one discussed in the next chapter) to physical tests of organoleptic properties.

This book is about food consumption behaviour and motivation, what people eat and why they eat this rather than something else; it is all based ultimately on consumer research. This chapter examines some of that research and asks how good it is and how it might be improved.

For a broad comparison between countries or over time (as in chapter 2) what people eat can be very broadly calculated from production and trade figures. For any detail, though, the only possible source of information is the consumer. Both the food balance approach and the relevant types of consumer research are examined.

In Part II, the 'determinants of food habits' were considered in separate chapters under the labels of economic, anthropological, sociological and psychological. However, in choosing among generic foods and among brands of those foods, the consumer somehow integrates all the different influences into single decisions. Research into the choices should thus start from those single decisions and work backwards to the separate influences. The second section of the chapter examines the methods used to research the ways we choose among foods.

Each chapter in Part III has its own main research methodology. Processing employs sensory testing which is a main element of consumer choice and is explored in chapter 11. Advertising is founded in the identification and measurement of what it is that influences the consumer, and uses much of the same methodology as Part II. Catering seems to rely largely on trial and error but does offer considerable opportunities for improving business decisions through the transfer to its activities of the research techniques developed elsewhere; new product development calls upon the whole battery of consumer behaviour and motivation. The research techniques are, again, those applied to answering the questions of what we eat and why.

Thus, the present chapter considers food consumer research under three headings:

(1) What food we buy and eat.
(2) Why we choose the foods we do choose.
(3) Of what foods do we like the taste, aroma, appearance (sensory testing), which is a subset of the reasons for our choices.

In examining the methods employed in the different types of research, the question arises of whether those methods are as powerful as they could be and whether they are usually applied as effectively as they should be. Disregarding odd mistakes, and such straightforward, if vital, matters as whether the data have been honestly collected and intelligently analysed, there are two conceptual reasons why market research may not be as good as it should be. They are:

(1) The decisions which depend on the research are not sufficiently specified.
(2) The research design does not adequately model the real-world situation about which the questions are being asked.

Taking the first point, effective marketing depends on accurate information. (This is a simplification; it depends on an accurate market assessment which might be based on information but might be the expression of informed guesswork.) The information is needed to help make better decisions—if it cannot be applied to a marketing decision, it is of no interest for marketing. Therefore, any market research must start from the specification of the decisions it is intended to inform. This is an absolute, invariable rule.

The second point arises from the fact that we are interested in the findings of the research only in so far as they tell us about the real world, the world in which the marketing decisions take effect. Therefore, the research is only useful if it reflects that world accurately—not in every detail, but in respect of the variables that are relevant to the decision. The research design must incorporate a selective abstract of the real world; it must model the real world. Otherwise the research is an irrelevance.

The chapter therefore ends with a brief comment on how research on the food consumer could be improved though observing these two requirements.

MEASUREMENT OF WHAT FOOD WE BUY AND EAT

The most basic question about our eating is '*What* do we eat?'

Commonly a government's aim in attempting to answer this question is to monitor the nourishment level of the population, or to provide information on price movements.

Disappearance Estimates

At the crudest level, nourishment is assessed through a food balance sheet. This is drawn up from statistics of production, stock changes, net trade, and diversion of food from consumption, to arrive at a total consumption estimate. The estimate can be divided by the population number to give average consumption per head per day.

Crude though the food balance sheet approach may be, it is used widely in North America to provide estimates of aggregate consumption trends for the major food commodities. Both the US Department of Agriculture and the Federal Ministry of Agriculture in Canada (Agriculture Canada) publish annual data on per capita disappearance of major food commodities in their respective countries. For example, calculation of per capita beef disappearance would entail:

— collect Federal and State/Provincial statistics on numbers of beef cattle slaughtered (including all grades of cattle—from choice steers to culled cows), estimating an average cold carcase weight for each category of cattle, and aggregating the total weight from each category to give a national total for domestically produced beef (in carcase weight equivalent units);

— add to the above figure, the total weight of beef imports in the given year. For each import category, an adjustment may be necessary to convert carcase equivalent weight,as much of the beef imported into North America is in boneless form;

— subtract the total weight of beef exports in the given year, again adjusting, if necessary, to a carcase weight equivalent basis;

— divide by the best estimate of total population;

and we have per capita disappearance of beef. The resulting statistic is in carcase weight equivalent units. It is an appropriate statistic for comparisons between countries or over time. For monitoring levels of nourishment, the quantities must be reduced to some estimate of cut-out values or retail yields.

Separate calculations on different foods lead to estimates of nutrient and calories per person, which in turn are compared with some assessment of desirable levels, and in some instances with medical examination of individuals. Of course, the uses of such figures are moot. Certainly, one could track per capita disappearance trends for the composite commodity category beef *vis-à-vis* pork; however, such estimates can provide no illumination on, say, consumption of beef sausages versus pork sausages, let alone consumption of competing brands of a particular type of sausage.

A food balance does not give any information on distribution among different consumers or even different sectors of consumers; this is a major shortcoming as these balances are most relevant in underdeveloped countries in which averages are of less immediate practical importance than the levels of income and food consumption per head of the most deprived members of the society.

Nor have food balances much commercial application. More pertinently for their own uses, firms in some subsectors of the food manufacturing and processing industries in Europe and North America can gain access, through their respective trade associations, to data on market size for certain food product categories. For many classes of foods, major manufacturers provide their trade association with sales/shipments information by product category. The association aggregates the figures and provides each member firm with data on national market size by product type (canned, dry, etc.). In turn, each member firm can compare its own sales with the industry total, and derive an estimate of market share, at least by product type. However, this information is still highly aggregated and such questions as who?/why?/where?/when? with respect to the purchase of a particular brand are still unanswered.

It is at this point that food consumption surveys are needed.

Consumption and Expenditure Surveys

In principle, the question of what is eaten is a relatively straightforward one to answer through market research. It is concerned solely with facts; there is none of the ambiguity or the arbitrariness of definition or the questionable validity of assumption that attaches to, say, the measurement of attitude. What we want to know is certain. We want to know what different subgroups of consumers eat at different times and places over the years. The decisions to be made on research design concern the choice of subgroups, the precision required, the degree of detail needed and so on. The difficulty lies in getting the information accurately and reliably.

Before discussing methods and problems of data gathering, it is worth indicating the relevance here of the point made earlier, that any survey or other measurement must start from a clear specification of what the findings are to be used for—of what choice of actions is to be based on those findings—otherwise there is no criterion by which to judge the precision or the detail needed. A researcher must be able to say, 'Because of the use to which the information is to be put, I need this degree of detail and this level of precision.' If he cannot say something of the sort, he is left with choosing arbitrarily or just doing whatever is convenient. The danger is particularly likely to appear with studies conducted by or for a government institution, which may commission them because of a diffuse belief that statistics should be collected and published, without any particular application being considered. The consequence is that the answers may be, for example, more detailed than the government might need for the creation of public policy, but less detailed than industry must have for arriving at commercial decisions. Admittedly, it seems a pretty tall order to demand that, before a government department (or anybody else) gathers a set of statistics, they must define what those statistics are to be used for. Who can say what someone at some future date will want? Yet, if that line of argument is carried far enough, every possible variable should be counted or measured and recorded in near infinite detail again and again and again, just in case. This is obvious nonsense. The only rational alternative is to attempt to visualize exactly what the data are to be or may be used for and then to make sure they are adequate for those uses.

To return to data collection, the difficulties are immediately obvious to anyone who asks himself the question, 'What did I eat last Wednesday?' Solutions, too, spring readily to mind. I do not think I can recall last Wednesday at all fully, but if asked about yesterday, I could, with a little prompting, probably manage to recall everything I ate or drank. Better still, had I been warned in advance, I could have written down each item at the time and presented a true record for yesterday, last Wednesday, or as long back as you would like. Of course, you would need to make it worth my while taking the trouble.

These, therefore, are the two methods commonly used to gather information on consumption or usage; data recall or a diary. Each has its limitations.

For recall the limitation is that demonstrated above. Even given great willingness to try, the period over which most of us can remember at all reliably how many cups of coffee we drank during the morning, or that we had eaten an apple between meals, is very short. Nevertheless, given a sufficiently short period, and taken systematically through it, prompted by an insistent interviewer, recall can be comprehensive. The method is flexible and will pick up items which, in spite of the instructions, tend not to be noted in a diary. Uncertainties or ambiguities can be resolved. Such questions as 'Is chewing-gum food?' can be answered.

The method is, however, expensive. The interviewer must be present and may have had to make an appointment, which involves a double contact. Because the information obtained will cover only a short period and relate only to the person interviewed, the sample will need to be large to get any substantial bank of data. Furthermore, because the respondent will not know exactly how much meat she had on her plate, the information must be incomplete.

The diary method is the obvious preferred one. It can conveniently cover the whole household and can record buying as well as eating, so that one measure is a check on the other and the results can be presented in terms of expenditure as well as consumption. Because the household is the unit of study, it is not necessary to know exactly how much meat each person ate: it is enough that, at whatever meals, the joint was disposed of. Once written, the data can be forgotten, so there is no need for detailed recall. The completed diary can be returned by post—or picked up in a brief visit, so that the method can be relatively inexpensive. As the same household can be used for more than one week, the demographic and other data about individuals and the whole household need be collected once only.

However, the diary method, too, is not without its disadvantages. It demands a lot of the respondent. Every little thing eaten must be recorded in detail. Anything eaten outside the home must be remembered. The method is therefore more accurate for food eaten at home than food eaten away from home. There is probably a tendency to underrecord items that attract social disapproval, such as sweets or alcoholic drinks. Because the method is so demanding of the respondent, some 'training' is needed. To make the method economical, therefore, each household must be kept in the sample for a period, which has its disadvantages as well as advantages; and because some households drop out before they are due to be retired, extras must be carried to ensure a balance in the sample. One weakness in the system which perhaps requires careful examination is the lack of direct control over when and how the diary is actually filled in; consequently, some households may throw together plausible entries when they are reminded to return the week's diary. The diary method is also unsuitable for an *ad hoc* study because of the time and expense involved in setting it up.

The two methods were compared in an experiment conducted by Stanton and Tucci (1982) in the USA. Observing that in-house marketing research departments often use surveys to save time and money, they compared data obtained through a 24-hour recall with that gained using diaries. They concluded that there was no difference in the results obtained by the two methods.

To treat this instance as a general conclusion would be rash; but, on the other hand, to undertake the series of large-scale surveys that would be necessary to confirm such a conclusion could be unjustifiably expensive. What would be practicable and very useful would be to match data gathered by the two methods from different commercial surveys where they are obtainable (for normally they are not published) as well as the publicly available results of a government survey, at different times and for different subgroups of the population. It could be argued that a comparison is not particularly useful since the two methods mostly have different applications, with recall being more often used for *ad hoc* information on individuals and diaries for continuing studies of households, so that only a limited proportion of the data would be common to the two sets of studies. Nevertheless, where the data were comparable, if the two methods could be shown to yield closely similar results, both could be considered validated. If they were shown to differ, the question would still remain of which was the more reliable.

National Surveys—the UK

Britain possesses the 'only long-term study of dietary patterns on a national scale anywhere in the world' (Frank *et al.*, 1984). It is the annual publication of *Household Food Consumption and Expenditure*—the report of the National Food Survey Committee. The committee, composed mainly of civil servants, but with some academics and business executives, is appointed by the Minister of Agriculture, and the work is funded by the ministry. The report is the common and universally quoted source of food purchase and consumption data for Great Britain.

The survey was begun during the Second World War. Its original form and purpose and the changes that have taken place are reviewed by Frank and her associates in the article mentioned above. Frank *et al.* also present a critical evaluation of the survey in its present form, making some of the points which follow here. The original purpose of the survey was to monitor food intake to ensure that the population (particularly the working class) was not undernourished during times when food was being imported at the cost of the alternative cargo of essential arms and ammunition and at the cost of lives of seamen. Today, the findings, arguably, have more economic than nutritional significance, offering analyses of consumption by socioeconomic groups, calculating elasticities of demand for different foods and so on. Its most significant use is in the calculation of the retail price index, providing the weightings for foodstuffs. The emphasis in the National Food Survey has shifted from its nutritional base; there is currently, however, a growing concern with nutrition and the effects on health of some of the more popular foods. Whilst this issue in the past has tended to concern only the medical and academic nuritionists, it may be that increasing public awareness and debate on preventive health measures through wise eating will awaken a relatively weak government interest and force official thinking. The National Food Survey, probably refined and extended, could again be a guide to policy making.

The survey uses the diary method. The details are succinctly described in an appendix to the annual report (MAFF., 1982).

> The National Food Survey is a continuous sampling enquiry into the domestic food consumption and expenditure of private households in Great Britain. Each household which participates in the Survey does so voluntarily, and without payment, for one week only. By regularly changing the households surveyed, information is obtained continuously throughout the year, except for a short break at Christmas and during General Election periods. Each household is provided with a specially designed log-book in which the housewife (or other nominee) records the description, quantity and—for purchases—the cost of food intended for human consumption which enters the household during the week it participates in the Survey. Ice-cream, fish and chips, and other take-away meals are excluded unless bought to eat in the home, and certain items which individual members of the family often purchase for themselves, such as chocolate, sugar confectionery, soft drinks, and alcohol are also excluded. Households are asked to record particulars of the number and type of meals obtained and consumed outside the home by each member of the family, but not of the cost or composition of such meals; however, the quantity of school milk obtained by children is recorded. To ensure that informants are recording food entries in sufficient detail, interviewers return to each household during and at the end of the Survey week to check the diaries. Information about characteristics of the household and of its members is recorded on a separate questionnaire.

The sample is a stratified, clustered, random sample of households in Great Britain. The actual sample size varies from year to year, but is about 7500 to 8000 households, out of some 14 500 households at the addresses originally selected. On the basis of a calculation reweighting different household groups in the 1971 survey to correspond with the census of that year, Derry and Buss (1984) conclude that 'any bias caused by non-response is probably not serious'.

As is evident, and acknowledged later in the appendix to the report quoted above, the survey measures food taken into households, not food consumed. It is thus, in its present form, gravely deficient as a measurement of nutrition levels. By not asking for details of, say, the occasional ice-cream eaten outside the home, it avoids one of the more intractable sources of inaccuracy, but it does so at the expense of completeness. By using each family for one week only, the survey seems to reduce the loss from households dropping out as they lose interest; and it saves having to pay them. Yet to get completed diaries from under 60 per cent of the originally selected sample offers the possibility of a very skewed sample. That the same household is not recorded at different times of the year must increase the sampling error; more precise results on the same sample size would be obtained were households paid for taking part and the same households used on more than one occasion—provided that the sample were representative in the first place.

Although the survey is not a true measure of food consumed, it can be a useful indicator of differences between regions, or between groups of respondents (say, by social class), and of changes over time. It can, however, do this only in so far as the causes of distortion operate in the same way across these differences. If the reasons why housewives in the higher social classes do not wish to take part are the same as those for the lower social classes (or at least, if the reasons do not reflect differences in food consumption) then the class comparison is probably unbiased—and perhaps this is so. But do the lower social classes eat proportionately more of their food away from home, or less of it? If the proportions are not the same, interclass comparisons will be distorted. And does the balance of what they eat away differ from home meals? If it does differ, home meals are not representative. Do we eat about the same proportion of our food out as we did 20 years ago? If not, what does the historical comparison show?

The survey sample is in the first instance selected for efficiency and representativeness, but it falls down on the response rate. The results are analysed with thoroughness and attention to niceties of statistical theory, and they are presented with clarity and due attention to caveats. But they are based on data that in important respects may be suspect.

Certainly, criticisms that can be made of one particular application of the diary method do not undermine the validity of the method itself. Rather they support the contention that if the data are not collected to meet the requirements of a precisely defined application, there is no criterion of adequacy in making the choice of data.[1]

National Surveys—North America

The USA cultivates a passion for publishing statistics, with the Department of Agriculture (USDA) second only to the United Nations as a source of agricultural and food data about all parts of the world. The USA has also shown an exceptional public interest in health and diet. It is, therefore, surprising that there is no continuing study of food consumption in the USA. What there is, instead, is a periodical Nationwide Food Consumption Survey carried out for the Consumer Nutrition Center of USDA.

The most recent US survey for which data are available is that of 1977/1978 (the previous one was undertaken in 1965/1966). The 1977/1978 survey was conducted over the period April 1977 to March 1978 and included some 38 000 individuals in the 48 coterminous states, using the method of 24-hour recall. The analysis covers patterns of eating and nutrient sources. The survey continues to spawn a flood of publications such as *Foods Commonly Eaten by Individuals: Amount per Quantity and per Eating Occasion*, by Pao *et al.* (1982). Other published analyses disaggregate the dollar expenditure on food into consumption and dietary levels of household by income, age group, region, etc. That the data on which such analyses are conducted are several years old, suggests an addiction more to history than to informed policy making.

In Canada, as in the USA, there is no continuing survey of food *consumption* although there are family *expenditure* surveys—which include family food expenditure—which are conducted from time to time (Statistics Canada, 1978). The vehicle for data collection on food expenditures is the diary. Quantities of food purchased for preparation at home are recorded, and data are classified by variables such as income level of head of household/family, family size, geographical region, etc. The primary stimulus and budgetary justification for collecting expenditure data is to monitor and periodically update and revise the consumer price index.

Besides the government-sponsored studies, there are, for most Western countries, research studies carried out by private market research agencies. Commonly these are paid for by clients who know what information they need for particular business decisions, and want it kept confidential, but in some instances the studies may result in published reports.

In the USA, the Market Research Corporation of America has repeated at 5-yearly intervals a National Household Menu Census. Over the period of one year, the census is conducted among 4000 households using a diary. Each household records every food item eaten by each member, and by any guests, over a fortnight. All items are recorded whether eaten in the house or away. The record is in full detail including brand and, for cooked dishes, all the ingredients used, as well as the method of preparation and appliances used. To ensure full reporting and to monitor the operation of the survey, the diaries are returned daily. At the end of the recording period, the participants respond to a battery of attitude questions. The sample is spread throughout the year. This would appear to be a thorough-going application of the diary technique.

Purchase Data

When the information on what people eat is gathered through diaries, it is common, almost necessary, to pick up information also on what they buy. As already pointed out, however, purchase behaviour over a short period is likely to be less representative of the larger period than is eating behaviour, because households eat regularly but buy food at different intervals which do not necessarily coincide with the data collection periods. The effect of spasmodic purchases will be removed if the sample is large enough and spread throughout the year. The size would tend to ensure than in any period households that bought more than they ate would be balanced by those that bought less; and the distribution through the year would cover the weeks, such as those before festivals, when households buy more than their weekly average, and the weeks in which they buy less. The difficulty, therefore, can be overcome straightforwardly by sufficient attention to sample size and distribution through the year; and spreading the sample over time is, anyway, more economical than covering it all at once.

This difficulty apart, purchase data recorded in diaries can be more precise than eating data. For one thing the housewife can simply check off her purchases as she unpacks them; for another, the exact quantities (and prices) are known. On the other hand, casual occasional purchases are as likely to be neglected as are casual occasional nibbles. There is, too, the same difficulty with purchases made by other members of the family; and if meals are bought outside the home, buying and eating are equally difficult to remember accurately.

More accurate and reliable data, though of a different nature, are gathered through the measurement of retail sales. Market research firms have long offered retail audit services to the grocery manufacturing sector. Manufacturers have been able to purchase information on volumes of specific products delivered to and moved from a representative sample of retail stores. The method has suffered the disadvantage of measuring the difference between what went into the store and what left, through whichever door. No distinction could be made between what was sold and what stolen or destroyed. (In so far as concern was with what was eaten by the households of customers and staff, this was a fairly satisfactory measure.) Yet, this form of retail audit has become a quaint anachronism with the advent of computerized retail checkouts and bar-coded grocery products. There is no practical reason why the details of every item that passes through the checkout should not be transmitted overnight to a central processor. Every detail and any required analysis can be on the decision maker's desk next morning.

Electronics do not, however, overcome one shortcoming of retail audits from the point of view of the manufacturer—indeed, it is accentuated. The shortcoming is that the very largest retailers may be unwilling to cooperate because their own sales give them a very good basis for themselves estimating total sales—so there would seem to be little advantage in pooling their figures to give information to competitors. Indeed, the pervasive introduction of computerization in the retail sector has engendered a fundamental shift in power in the consumer research market place; major retail chains have up-to-the-minute access to research data that are hidden from and of great value to market research agencies and their manufacturer clients. However, the great advance is that the information is rapidly and readily available to the retailer. What bargain the research agency or manufacturer must strike to gain access to the information is quite another matter.

Its direct access to the consumer gives the retailer a research advantage which a manufacturer cannot match. The logical consequence (and it is already manifested in North America) is for the manufacturer to buy direct from the retailer information on sales (and, eventually, consumption) of his own products and those of his competitors.

Computer-generated retail audit data do not directly indicate who buys what. However, there is no difficulty about extending the information gathering to cover consumer data. Already stores which offer their own credit cards have access to the shopper's home address and some knowledge of his financial status. More can readily be learned. By setting up its own panel of shoppers, representative of the population in the area, and by giving panel members sufficient incentive to buy their groceries normally from its stores, a retail group could (should it wish to) produce day-by-day purchase information as accurate and detailed as that yielded by national surveys. The shopper panel could also be induced to keep diaries recording consumption, providing the full set of data and, indeed, acting as a double check on recording accuracy of at least the store purchases.

Catering Data

Publicly available market data on food consumed away from the home are few and far between, both in Europe and North America. However, the National Restaurant Association (NRA) in the USA does publish, periodically, trends in food consumed away from home (the CREST reports), classified by restaurant type; for example, 'take-out' food outlets by major meal category (pizza, hamburger, fish, etc.), full-service family restaurants, steak houses. Member firms of the NRA provide information on their food outlets by location and meal type, and surveys of customers are published that identify customer demographics and average cheque size, etc.

In the UK, the National Food Survey provides an estimate of meals eaten outside of the home, analysed by region and income group. This special analysis is based on information in the diaries regarding absence from meal occasions at home by members of the household.

To obtain information on the breakdown of these meals by outlet, expenditure, percentage of meal occasions accounted for by meals eaten outside of the home, we have to turn to commercial sources, notably Gallup, who provide the 'British Eating Out Survey' based on a 2500 sample using a diary system. To encourage participation, a reward incentive, in the form of a premium bond 'draw', is offered. Like the National Food Survey, this survey fails to record information on ice-cream and confectionery bought on a 'casual' basis. Other major sources of information are provided by Keynote and Euromonitor publications, but again, emphasis seems to be on economic as opposed to nutritional aspects; understandably so, considering their commercial application.

As the number of meals eaten outside of the home increases, with fewer family meal occasions and changing lifestyles, the policy makers need to be aware of these changes and their dietary implications.

There are, of course, on both sides of the Atlantic, *ad hoc* studies conducted for particular applications; but the findings are generally not published.

MEASUREMENT OF REASONS FOR CHOICE

The question of why people choose the foods they do is barely less fulsomely researched than that of what they choose. This question, however, is asked almost solely by commercial firms, mainly manufacturing firms. Their object is to influence choice; it is in the clear interest of competing firms to discover ways in which their products could be made more appealing to their customers or to potential customers.

The attitude of governments in Western countries seems largely to be that, because there is no real danger of undernutrition, it is not their function to interfere with consumer choice. Nevertheless, it is now beginning to impinge on the public consciousness that, though there is no general problem of undernutrition, there is a real and menacing problem of malnutrition. If any part of the purpose of eating is to sustain life (and, however food choice may be overheaped with other subtler, less fundamental but yet dominating motivations, we do eat to live), then many of us have reached the self-defeating state in which what we eat may be killing us. Here, where public interest may seem to clash with the private advantage of producers, governments will surely become more responsive to public unease, and are very likely indeed to begin to want to influence consumer choice or at least the development of habits that condition the choice. They, too, will become sponsors of research into why consumers choose the foods they do.

In less developed countries, the situation is different. In many such countries, there is starkly obvious undernutrition and, in some, starvation. For large sections of the populations in these countries there is small need to discover why consumers choose between foods. Government preoccupation must be with getting enough food to the populace rather than influencing what food people choose; and problems of malnutrition arise because the right foods are not obtainable rather than because consumers prefer the wrong ones. In lands or among sectors of the population where the situation is not so desperate, there are, however, even in less developed countries, dangers of malnutrition through choice of the wrong foods.

The attempts to answer the question of why consumers choose particular foods are most commonly predicated on the assumption that attitudes condition behaviour, so that to measure attitude is to measure how people are likely to behave, and thus (with a small leap of logic) to uncover the reasons for this behaviour. This appears to oversimplify or even ignore the complexities of the interrelationship between attitude and behaviour, but does not really do so. On the definition of attitude commonly used in marketing, of 'a predisposition to behave', what is being attempted is to measure the factors, cognitive and emotive, which influence a particular class of behaviour; and, since the whole purpose of the measurement is to discover how behaviour can be influenced, nothing more is needed. The techniques used are those of attitude measurement.

A more satisfactory presentation of the analysis starts from the definition of the product as a bundle of perceived want satisfactions, physical and psychological. This, which has long been a commonplace of marketing analysis, was introduced to economists by Lancaster (1971) in his 'characteristics of goods' theory, in which he argued that consumers do not buy products but the attributes of those products. In fact, it is not actual attributes in any objective sense, but the consumer's perception of those attributes in relation to satisfying his or her own perceived wants. The object is to measure these perceptions.

The standard approach is through an exploratory stage to establish the range of perceptions, followed by a quantifying stage to measure both the importance of each dimension of perception and the rating of each food or brand on that dimension.

Easily the most common way of carrying out the exploratory work is through focus group discussions or unstructured intensive interviews with individuals. The focus group generally is made up of about eight people, most often of similar enough background or interests to be able to discuss comfortably whatever is under investigation. The group lasts an hour and a half or so; respondents are paid a small amount and offered light refreshments. A moderator, working to a schedule of the topics to be covered, leads the group through the schedule as unobtrusively as possible, encouraging discussion among the group rather than dialogues with himself or herself. The discussion is recorded and later analysed. Although some groups prove more fruitful than others, a skilled moderator should be able to educe from the group at least a perspective on what issues or dimensions are relevant to the research questions.

The alternative to focus group discussions is fairly unstructured interviews with one person at a time. They have the advantage of being able to penetrate to a greater depth if conducted with enough skill, but, of course, they lack the group's advantage that one person's comments can spark off someone else.

The discussion or interview often includes some projective tests, many of which are watered down versions of the procedures of clinical psychology. Their purpose is to make it easier for respondents to express opinions or feelings they may be inhibited from revealing or are not able to articulate readily or of which they are not overtly conscious. Examples of the techniques are: Thematic Apperception Tests (TAT) in which the respondent is shown a picture in which the issues being researched are left undefined (e.g. the brand of product) and he or she is asked to describe or tell a story about what is happening; 'balloon' tests in which a respondent is asked to fill in conversation in a cartoon-like 'balloon' issuing from a speaker's mouth; word association and sentence completion; role playing; personification of a product.

The quantification stage usually relies mainly on one or more of the many different types of rating scale. The scale will offer the respondent a set of numbers or words or symbols, usually in some progressive order, and she will be asked to rate some product or whatever on the scale. A simple version is 'Give each of these apples marks out of 10 for sweetness'. The two most frequently used are the semantic differential and the 'agree-disagree' types. An example of the semantic differential is

High-priced								Low-priced

The respondent is asked to tick the box which best represents her opinion of the price level of each of the products studied. In the 'agree-disagree' type, she is confronted with a statement such as 'Apples are good for you' and asked to indicate her agreement or disagreement on a scale like:

Agree very strongly	Agree strongly	Agree slightly	Neither agree nor disagree	Disagree slightly	Disagree strongly	Disagree very strongly

There is a very wide range of possible scales. Those shown above, for instance, are both seven-point scales, but there could be anything from two points upwards. Instead of words there could be numbers, or degrees of shading or other symbols. They could be presented horizontally, as here, or vertically or in some other formation. They could, as here, allow the respondent to award as many marks or points as she wishes to each product, or she might be forced to choose by being given a fixed number of points to allocate among products. These rating scales do, like the projective tests, help the respondent to express comments which otherwise she could not do. Because these scales can be included in questionnaires and can be administered quite rapidly, there is no limitation on the size of sample, and a fairly high limit on the number of objects which can be rated. Consequently, large batteries of such scales are administered to samples of sometimes upwards of a thousand respondents.

The findings are analysed by one or more multidimensional techniques, e.g. the analysis can (as in factor or principal component analyses) show underlying relationships among the scales or among the objects scaled, and the relationships of products to some ideal product; or (as in cluster or discriminant analyses) it can identify relatively homogenous groups of consumers and their common perceptions; or it can examine some other set of relationships in the data. In fact, the statistical techniques available can offer answers to most of the questions asked by anyone wanting to influence consumer choice.

These methods have been very widely used for many years. Do they, however, yield valid answers? It is unlikely that they will reveal the whole truth, but is it enough of the whole truth to lead to effective action and as much as could reasonably be discovered for the work and expense involved? In practice, it must be doubted whether the answers are always the best that could be obtained, because of the many elements in the whole process that may be either fudged or left to arbitrary judgement. A brief examination of the process shows up the problems.

To start at the beginning, this book has five chapters concerned with separate influences on food choice. The first is on nutrition, and there is a chapter each on the economic, the anthropological, the sociological and the psychological influences on eating and food choice. Some of these issues are touched upon again, and some further ones introduced, in the present part of the book. It seems a lot for one consumer to take into account when deciding to buy or eat a can of beans.

Of course, it does not quite work like that. The different influences are brought to bear in different ways at different times. They generate perceptions of the consumer's self and her needs, and of particular foods, as commodities and brands, in relation to the needs. The combined result is a predisposition to buy or to eat, which at some moment is prompted by some cue to become the decision to buy or eat and the actual buying or eating. Even during the process of buying or, yet more so, of eating, the perceptions of need and foodstuff change further. These perceptions may be relatively stable, but they are dynamic.

Naturally, the buyer or consumer is not conscious of all this. Some of the perceptions may be precisely formulated and deliberately considered. In most Western countries, the price and nett weight of each item of food is clearly marked on the pack. The buyer is provided with the information needed for comparison between two offers, although the precise calculation, through different sizes and prices, is likely to defeat any but the most systematic shopper armed with pencil, pad and pocket calculator; most shoppers probably consider life too short to bother. Yet almost every buyer will have some general awareness of the price level of most of the foods she buys frequently; the awareness may be peripheral, in that it could be recalled but is not in the forefront of her mind, but if some price went dramatically out of line she would notice. And, because the only standards of high or low price are comparative, the buyer must relate each price (perhaps casually or perhaps with careful calculation) to competing prices and to the money at her disposal. The perceived price will have an effect on this and future decisions.

Other attributes, such as some element of the taste, cannot be so clearly and objectively specified; consequently, the experience cannot be so definitely codified and therefore cannot be so precisely recorded or remembered. But even those perceptions, which may be more of feeling than of cognition, will have a direct influence on present and future behaviour; indeed, their influence can be overwhelming. And it is this sort of perception that is changing all the time: how much someone likes butter adapts each time butter or a substitute is eaten. If the perceptions are buried below the surface of consciousness, their origins must be even further obscured from the consumers themselves.

Yet, although the consumer may not know or be concerned about exactly what influences underlie her behaviour, the researcher who sets out to show how the behaviour can be changed must be able to disentangle this network of influences and to identify those that are important, in order to hypothesize how they operate and how they can be interfered with precisely and accurately. It is to this task that the research methods need to be equal. Are they?

The answer must be, at best, 'perhaps'! A skilfully conducted focus group, or, rather, a sequence of well-planned and skilfully conducted focus groups, may expose the whole range of the relevant factors, but there is no way of being sure they have done so. Conventional wisdom, founded on experience, has it that after six or eight groups in a not too diverse population, a researcher is unlikely to learn any more by adding further groups. However, the fact that further groups will not reveal anything new does not necessarily show there is nothing further or deeper to be uncovered; it may indicate no more than that on the topic in question further groups will not go any deeper than the groups already conducted. It is a common enough experience for the focus group to tend to confirm what is already known, to add a few new angles; possibly, to introduce one or two quite novel ideas. But is there more?

To some extent the projective techniques of the early days of 'motivation research' have fallen into disuse. Some practitioners make them a main technique of their research, but most seem to include them indifferently within focus groups. Certainly, the focus group is a proper habitat for projective techniques; but were tailor-made projective techniques a central part of the focus group sequence, qualitative or exploratory researchers might strike richer seams of consumer perception.

The quantifying methods invite even stronger scepticism. If a housewife is asked to place in order of importance the attributes by which she chooses food, is she likely to put 'the look of it' very high on the list? Will not nourishment, taste, price and convenience come higher? Yet in so far as the cook gets credit for the presentation of the food, and in so far a the housewife hopes to get credit from her cooking, she must rate its appearance highly. Women's magazines, with their enticing pictures of food and a good deal of emphasis on culinary decoration, clearly do attach importance to the look of food. But attention to that apparently incidental characteristic of a dish might seem frivolous, and who wants to label oneself so? If appearance is not rated as important, the reason may be the respondent's desire to appear intelligent and serious, rather than any indifference to the presentation of the food. Are not people prone to respond in the way they expect to be approved?

Furthermore, presented with a few pages of scales, does a respondent try throughout to reflect, in her marking, her true attitude, or does she at some stage change to getting through as quickly as she can? In practice it is the interviewer not the researcher who is in a position to judge whether the respondent gives up trying, and the interviewer, too, has an interest in getting to the end of the questionnaire. Specific research studies to measure the attention span of a respondent must suffer from the drawback that the span will vary with the respondent's interest in the subject, so that generalization is limited. Nevertheless, the question remains.

The analytical methods, too, raise doubts. The calculations are swift and accurate, but how many users of multidimensional statistical analyses fully consider or even comprehend the assumptions that underlie the connection between the calculation and the conclusion? One of the simplest criticisms lies in the treatment of ordinal scales as interval ones: clearly, 'agree strongly' denotes more agreement than 'agree' so that the markings on the scales are certainly ordinal, but is there the same distance—in the minds of all respondents—between 'agree' and 'agree strongly' as there is between 'agree strongly' and 'agree very strongly'? Yet as, more often than not, the analyses used require interval scales, the assumption is being made, even if not explicated, that the intervals are equal.

At the very last, therefore, the answers yielded by the research are open to question.

ASSESSING THE PHYSICAL PRODUCT

A whole world of food consumer research is that of sensory testing. The practitioners of this activity would probably claim that theirs is the most scientific element in food research. Its methods and instruments are closely related to those of the natural sciences, and many of the researchers have themselves been trained in one or other of the natural science disciplines. In Britain the professional home of sensory testers is the Sensory Panel of the Society of Chemical Industry's Food Group. Contrariwise, it can be argued that this is the least scientific part of food research, because in spite of its trappings, more frequently than in other food research, sensory testing neglects to test what is relevant to the marketing decisions to be made, but, often with considerable ingenuity, manages to test something else instead.

What sensory testing sets out to measure is the physical acceptability of the product to the consumer, that is the acceptability of the physical attributes of the product as perceived directly through the senses, the attributes of appearance, smell, texture, taste and perhaps after-effects. The problem, which possibly affects sensory testing more than other forms of market research, is that identified earlier, the problem of modelling. What is being modelled is the perceptions of ordinary consumers encountering the product as it is presented to them in supermarkets or on the table, often after they have been exposed to advertising. The test itself will, in many instances, be conducted in a laboratory or under 'laboratory conditions'. Probably, most natural scientists are accustomed to carrying out experiments in laboratories; and laboratory conditions lend themselves to controlled experimentation in which you know what you are testing. It is, however, impossible to reproduce in a laboratory the conditions in which, in real life, the consumer will try the product. Sensory testers do recognize the problem (see, for example, Sidel *et al.*, 1983), but a great part of all sensory testing still goes on in laboratories. Furthermore, attempts are made to produce instrumental measures which correlate satisfactorily with consumer ratings—when the 'consumers' are commonly panels of experienced testers operating under laboratory conditions.

It may indeed be impossible to model adequately the conditions in which consumers make up their minds, perhaps often trying the product more than once and usually being subject to the pressures of promotion, image creation, comments of friends, price and other circumstances. Yet these are the relevant conditions. If a product is to succeed it must be accepted in such conditions, and unless the test can come near enough to those conditions to act as a predictor of what would happen in them, it is useless, even actively misleading.

Land (1983) distinguishes four classes of sensory test: (1) detectability of some difference; (2) the intensity of that difference; (3) the quality, the character or description of the attributes present; and (4) reactions or liking of the product. This is a useful classification as it progresses from the detectability of a difference to the liking of the product, though it could be extended to a fifth level of what it is that is liked about that product. What is vital, though, is that the tests are related to the target consumers. It has been argued (see, for example, Lesser, 1983) that if the consumer cannot detect a difference then there is no difference in the context of marketing, allowing for the fact that she may 'detect' a difference not from the physical product but because of what she reads on the side of the pack. Conversely, if by every other test two products can be shown to be identical, but the consumer—or enough consumers—believe they can detect a difference, then the products *are* different. The only marketing reality is the aggregate of consumer perceptions and it is this perceived difference which must be revealed.

Much sensory testing uses trained panellists to detect differences—Land's first class of test. Their use is justifiable solely on the argument that, if these more sensitive consumers cannot detect a difference, then the rest of us cannot. If these panellists *can* detect a difference, then to have any practical applicability the test must be repeated on representative consumers. Of the other classes of test distinguished by Land, expert panels have some application in measuring the intensity of a difference and in describing product attributes, but only as leads for later tests on less sensitive consumers. When it comes to reaction or preferences, trained panellists have no place. The tests must be conducted on 'ordinary' consumers in conditions that adequately reproduce 'ordinary' conditions.

There is no especial difficulty about finding 'ordinary consumers'; for any product they are the defined target group, and a representative sample of them can be recruited as testers. 'Ordinary conditions' are rather more difficult to establish. Generally, the testers will be given the product free or sold at a reduced price; and they will know they are testing—they will be more self-conscious than is normal. The bias introduced by these abnormalities, however, may not be very great and can be countered in the experimental design.

What cannot be fully simulated is the consumer's creeping awareness of the product through its promotion, the conditioning of expectations which is likely to precede the physical experience of the product. In some cases, possibly most, they may not *seem* to matter: if two products are being compared, the conditions for the two are the same, and it can be argued that the less the result is affected by external influences, the better the test is of the physical characteristics alone—the fewer the variables, the better controlled the experiment. The argument is specious. Certainly, the two products should be compared under likc conditions, but thcsc must rcproducc,in cvcry way that would make a difference to the choice, the conditions under which the product will be encountered in real life. Sensory preferences are not objective or independent of other influences. We learn tastes, and we learn them in respect of circumstances; a taste that delights in a cheese may be disgusting in an egg. There is no escaping the need for verisimilitude.

Given the representative sample of consumers in surroundings that are a valid reproduction of reality, there is no particular difficulty in getting respondents to rank two or more products in order of preference. Product preferences are, however, seldom unidimensional. A consumer may rate the appearance of one orange superior to that of another but the taste inferior. An overall preference rating will simply conceal these counterbalances, or rather will subsume them through giving different weights to different variables and applying the weights to the measurement differences. Thus, the appearance of the first orange may be very much better, while the taste of the second is only slightly preferred; yet because taste is regarded as much more important (and so given a much greater weighting) the overall preference could go to the second.

The analysis must extend much further. Appearance is not unidimensional. It comprises colour, shape, smoothness, reflectivity and more. Taste is similarly multidimensional, as are other sensory attributes. The research, pursuing products that will appeal more, must unravel all the different strands, so that each may be adjusted individually, for that is all that can sensibly be done. The alternative is to alter different elements at random, checking each time for increased or reduced overall preference; which may work but is more likely to end in sheer frustration.

Thus, the first step is to identify all the different elements of the different sensory attributes that contribute to the consumer's perceptions of the physical product. The second is to get the respondent to rate each product on each attribute in appropriate circumstances. The third is to arrive at the weightings for each element that accurately reflect the weightings imposed, consciously or otherwise, by the consumer. The fourth is to combine all the weighted elements into a set of notional products: from the first three stages it will be possible to calculate the ratings of overall preference that consumers should give to each of the notional products; if these ratings match the ratings of overall preference that consumers actually do give to the real products, then the different elements of the first three stages will have been validated. The whole scheme of sensory tests will have been proven.

GETTING THE RESEARCH RIGHT

How, then, does the market research practitioner ensure that his research is sound? The standard he must demand is two-fold:

(1) The research must be more cost-effective than guessing: it must improve the decision by more than it costs. There are straightforward ways of calculating the probable payoff from undertaking research, and the cost if ascertainable in advance; so, given clear thinking, this condition presents no problems.
(2) It must be as accurate and as precise as it can be made for the cost. This is obvious. The only further question is how it can be made so.

The attempt to ensure the quality of research or to improve it must be based on what is likely to be deficient. The available research tools are broadly adequate; it is mainly in their application that fault may lie. The two major causes of shortcomings in research practice were earlier suggested to be insufficient attention to the relevant marketing decisions and inadequate modelling of the real world. To remedy these deficiencies, where they exist, the researcher must therefore:

(1) Specify fully the decisions the research is intended to inform and design it to do just that and nothing more.
(2) Ensure that the design does reproduce adequately the elements of the real world that are relevant to the decision—that the model is accurate.

About the first condition there is little more to be said, except perhaps to repeat that the only starting point for market research is a set of particular decisions that need to be made: unless these are clearly identified and motivate every element of the research process, that process can only be sound by chance.

The second point can bear more elaboration. The argument goes thus. In the end all marketing decisions rely upon and relate to some model of the market and the processes that induce change in that market. The model specifies the variables that have an important bearing on the decision and postulates the relationships among them. Every marketeer must say to himself (or at least implicitly accept), 'This is what I think (or feel) will happen. This is what I think people want the product for; this is what I think they will do with it; this is why I think they will buy it.' He is establishing his model of the market.

If the model is accurate, and if he then acts appropriately in consideration of that model, the product will succeed. If the model is wrong, the product will be wrong (unless by chance it happens to fit the different real market). In practice, and in spite of all the courses on business decision making, the model may not be systematically derived or precisely explicated. It may just be arrived at intuitively and remain an individual's feeling—he 'knows' something will work: still, if it is accurate the product will be right. Indeed, the success of natural marketeers, of people who have got it right, must rest on their knowing what the market wants, on their having constructed intuitively an accurate model of the market. They do not need market research. The rest of us do.

The process works like this. The marketeer hypothesizes a model. It must be specified in detail. It relates to consumer perceptions of wants and of products as want satisfactions. If the consumer's behaviour, as a consumer or as a purchaser, is to be changed, her perceptions of her wants or of the possible or preferred want satisfactions must be changed. The model must therefore cover the relevant perceptions and also the channels of influence through which the perceptions can be affected. The hypothetical model is thus a specification of assumptions and of the implications of those assumptions. Most of these assumptions can be accepted without testing: there is a risk that one or another is wrong, but the risk is estimated to be sufficiently small as to cause no further concern. Some of the assumptions are, however, sufficiently important, and there is enough uncertainty about them, for it to be necessary to test them by research. Even the research carries the risk that it could yield the wrong answer, but the risk is lower than that of guesswork.

For example, a product needs to be named and it can be accepted that the name will have a significant bearing on the likely success or failure of the product. In a particular instance it may be that a name has been thought of that seems an utter winner. There would then be little point in going through an elaborate programme of name testing—some brief checking that it carries no negative connotations in any of the proposed markets will probably suffice. The same balance of benefit against cost will take place for every element of the 'marketing mix'. On another level, a product may so obviously meet a long-identified want as to give an almost certain assurance of success; in other cases the fundamental question is whether or not the concept has sufficient appeal to be worthwhile.

The basic point, therefore, is that failure, where it occurs, is less likely to be brought about by inherent weakness in available research techniques than by inadequate modelling and therefore improper specification of the questions research is needed to answer. This may seem a handy get-out. 'Don't blame the researcher, blame the marketeer; it's the fault not of the research contractor, but the client.' Yet it must be valid. A marketeer who knows what questions he wants answered should also know whether any particular piece of research is likely to answer those questions.

Arriving at a model, sufficiently detailed and yet not cluttered with irrelevance, is admittedly a major challenge. There are a number of general models of consumer decision making. Examples are those of Engel *et al.* (1968), Howard and Sheth (1969) and Andreasen (1965). They may be broadly applicable in any instance, and attempting to fit any one of them, or some amalgam, to the particular product market is a healthy beginning. The end, though, needs to be much more refined: each marketeer and, more especially, each would-be developer of a new product must specify a particular model of the market for that very product and of the influence processes that operate in that market. That model must be the starting point of his decision making and his research. If it is an accurate model and he applies it wisely, his research should be beyond criticism.

To return briefly to the types of research discussed earlier, consider how they might be improved in the light of the points made here.

Both the main methods of researching what people buy and eat present difficulties in collecting complete data. These can be overcome by painstaking and persistent attention to detail. However, knowing how much detail is needed and with what degree of precision (indeed, what data should be collected at all) can only follow from knowing what it is to be used for. It is broadly true that, in this respect, government-sponsored surveys are lacking, commercially sponsored ones satisfactory.

On why we choose the foods we do, the methods of exploratory research are so clearly an attempt to model the real world—to discover how consumers perceive it—that they are unlikely to fall down on that point. There is rather more danger that, because the approach needs to be openminded, it may be thought that the subsequent marketing decisions need not be specified in advance. They must so be, or else what is to be explored and what not? As 'anon' has commented, 'An open mind need not be an empty mind.' In fact, however, the greatest difficulty in exploratory research is one of technique, of ensuring that the 'exploration' does penetrate to the true perceptions rather than to commonly articulated ones.

At the quantification stage of attitude research, too, many of the more obvious failings are technical ones: the invalid use of a multiplicity of scales; the confusion of ordinal and interval measurement; and the inappropriate use of statistical analyses. Underneath, though, these also are problems of modelling. Does the scale enable the respondent to project and express adequately her real feelings or opinions? Do the assumptions of the analysis match the real world?

On sensory testing, deficiencies of modelling are the paramount fault (see Lesser, 1983). As scaling is a basic technique of these tests, some of the work is open to the same criticisms as those made above of the quantification stage of attitude research; but, again, even here the underlying problems are those of modelling. Far more obvious, though, are the difficulties of reproducing adequately in the laboratory the conditions of the real world—of devising an adequate model.

In summary, therefore, the foremost requirement of good market research is relevance: it must be tightly relevant to specified marketing decisions; and it must reproduce the elements of the real world to which the decisions relate. Even given these two requirements there are innumerable ways in which research can fail. However, many of these findings can be traced back to a failure to satisfy this double test of relevance, and any 'research' that does not satisfy the test cannot really qualify as market research.

1. It should be said that the Committee of the National Food Survey (of which one of the editors is a member) is well aware of the limitation imposed by the Survey omitting food eaten outside of the home, and is actively considering ways of overcoming this limitation.

REFERENCES

Andreasen, Alan R. (1965) 'Attitudes and consumer behaviour: a decision model'. In Preston, Lee E. (ed.) *New Research in Marketing*, Institute of Business and Economic Research, University of California, Berkeley.

Derry, B. J., and Buss, D. H. (1984) 'The British National Food Survey as a major epidemiological resource', *British Medical Journal*, March, **288**, 765-767.

Engel, James F., Kollat, David J., and Blackwell, Roger D. (1968) *Consumer Behaviour*. Holt, Rinehart & Winston, New York.

Frank, Judith, D., Fallows, Stephen J., and Wheelock, J. Verner (1984) 'Britain's National Food Survey—whose purpose does it serve?' *Food Policy*, February, 53-67

Howard, John A., and Sheth, Jagdish N. (1969) *The Theory of Buyer Behaviour*, Wiley, New York.

Lancaster, Kelvin (1971) *Consumer Demand—A New Approach*. Columbia University Press, New York.

Land, D. G. (1983) 'What is sensory quality'. In *Sensory Quality in Foods and Beverages*, ed. A. A. Williams and R. K. Atkin. Ellis Horwood, Chichester. 15-29.

Lesser, D. (1983) 'Marketing and sensory quality'. In *Sensory Quality in Foods and Beverages*, ed. A. A. Williams and R. K. Atkin. Ellis Horwood, Chichester. 448-466.

MAFF (Ministry of Agriculture, Fisheries and Food) (1982) Appendix A. *Household Food Consumption and Expenditure 1980*. Annual report of the National Food Survey Committee, HMSO, London.

Pao, Elizabeth M., Flemming, Kathryn, Greneker, Patricia, and Mickle, Sharon (1982) *'Foods commonly eaten by individuals: amount per quantity and for eating occasion'*. Home Economics Research Report no. 44, US Department of Agriculture, March.

Sidel, J. L., Stone, H., and Bloomquist, J. 'Industrial approaches to defining quality'. In *Sensory Quality in Foods and Beverages*, ed. A. A. Williams and R. K. Atkin. Ellis Horwood, Chichester. 48-58

Stanton, J. L., and Tucci, L. A. (1982) 'The measurement of consumption: a comparison of surveys and diaries', *Journal of Marketing Research*, May, 274-277

Statistics Canada (1978) 'Urban family food expenditure, 1978', Statistics Canada, Catalogue 62-548 Occasional.

The Food Consumer
Edited by C. Ritson, L. Gofton and J. McKenzie

CHAPTER 10

New Product Development: The Role of Social Change Analysis

ELIZABETH NELSON

SOCIAL CHANGE AND THE MARKETING FUNCTION

The consumer boom of the postwar period brought with it a new development in company thinking—the discovery of the marketing function. As awareness of the importance of product targeting grew, so did the amount of time spent on investigating how consumers reacted to new products.

Thirty years on, the marketing function in many companies has reached a highly sophisticated level. The marketing function assumes a strategic importance in many companies and some leading corporations incorporate into this an analysis of changing social values.

Traditionally, new product development was the sole responsibility of either the marketing *or* the research and development department; only occasionally was it a joint responsibility. Recently, the most successful firms have come to view new product development in the light of how society is changing. They recognize that, because social changes have been so profound and dramatic over the last two decades, they have had to alter management structure to allow for greater flexibility in responding to consumer needs.

For many people in all societies in the Western world, the desire to be respected by others, and to respect oneself, predominates and governs much of their behaviour. Social movements have shifted away from displaying one's wealth by appearing to be well-off, and better-off than others. The quest for status has diminished; hence the desire to be *seen* to own the latest and to be an innovator has lessened.

During the 1960s and 1970s, the status motive lost its force. Consumer behaviour is now motivated by desires related to quality of life, wellbeing and self-expression. This does not mean to say, however, that there are not some people who still feel a need to announce their status and success by means of conspicuous behaviour. We have examples in the UK of people who purchase certain important consumer durables, such as coffee filters, microwave ovens or expensive cars, because they wish to demonstrate their consumer spending ability to others. But increasingly, the predominate feeling is guilt about showing off status and about ostentation. People will increasingly buy expensive consumer durables claiming that they are not for prestige or status, but for things such as quality, comfort and safety.

The marketing function in management was originally built upon tenets of conspicuous consumption, modernism and a systematic attraction to novelty. The successful corporations recognize that these have now begun to deteriorate. People become more open to change, yet begin to attach greater value to old things, and to question innovation for the sake of innovation.

The successful marketing company recognizes that the break-up of the status system represents a threat to clearly hierarchic ranges and it offers an opportunity for new types of ranges and products. Traditionally, the marketing function was very rigid. Managers, armed with the market research and econometric methods, could relatively easily predict consumption and lifestyle patterns. Now the best placed corporations are those which are capable of coming to terms with the ideas that the future is not predictable and that society is heterogeneous. They are, therefore, capable of devising new methods and new products suited to the present-day circumstances.

The corporations best placed to develop new ideas are those which are capable of changing their view of the environment, seeing it as more variable, more changeable, more complex, and which are alert for straws in the wind. These are but a handful of sophisticated marketing corporations.

A gap has arisen between the growing *need* for products and services (adapted to the expectations of individuals) and supply which has remained standardized and based on mass-production. In certain sectors, consumer aspirations and sensitivities have changed, while goods and supply have remained unmodified or are even evolving in a different direction.

In Table 10.1 data are shown from the MONITOR surveys (see Appendix I for technical details of MONITOR) which illustrate this gap. Large minorities of the population indicate that in many product fields they cannot find models or types which entirely suit them.

At the time of fieldwork, TV ratings were declining steadily, and it is interesting to note that less than one-third of the sample said that TV programmes entirely suited them—a finding repeated in similar social trend surveys throughout Europe.

If we examine in greater depth the products/services deficit, we find that here are vast opportunities for both product and social innovation. The core social trends most conducive to these deficits are related to self-expression.

Let us outline some of the main social changes observed in the last two decades.

Table 10.1 Product deficits.

% respondents who indicated that 'All that I want is available in...'	
Consumer appliances	76
Bread	75
Shampoo	71
Clothes	68
Breakfast cereals	68
Medicines one can buy for oneself	65
Frozen foods	64
Non-alcoholic drinks	62
Confectionery	57
Cars	54
Snack products	49
Alcoholic drinks	47

Based on a random sample of 1533 respondents: Taylor Nelson Monitor Survey 1983.

(1) A growing desire for personal autonomy or independence.
(2) Fewer differences between the attitudes of women and men—a blurring of sexual identity.
(3) An increasing acceptance of a less structured or organized way of living.

A psychologist could interpret these changes as signs of a broad movement towards an 'inner-directed society', i.e. a society where people are more concerned about the individual and creative aspects of their existence than about status and conformity. I would not dispute this but would point out that society is capable of moving in a number of directions and that there is no one clear path pointing to the future. Inner-directed values coexist (sometimes uneasily) with the status-conscious outer-directed ones; and, of course, there are still a considerable number of people whose primary concern is to keep their head above water (and as such could be called 'sustenance directed').

Although society may be moving in a number of different directions, there is a clear relationship between the three core trends: Personal Autonomy, Equal Opportunities for Women, Informality.

All inner-directed trends, and feelings that what is available does not meet consumer needs, help to isolate areas for new product development.

USING SOCIAL TRENDS AS A FRAMEWORK IN NEW PRODUCT INNOVATION

One can use the social trends, and data on who is exhibiting the trends, as a framework for marketing. The dynamics of society force managers to think in terms of growing trends. With social change analysis, operating managers are offered a framework which enables them to see threats and opportunities and to create new ideas which can exploit the growing social trends.

To return to product deficits, and taking the two areas 'alcoholic drinks' and 'snack products', we find that several social trends are highly correlated with the gap, as shown in Table 10.2. Take, for example, the trend Informality. It is made up of four attitude statements. The sum of the scores across all four items is calculated for each of the 1533 respondents in the 1983 sample. We are particularly interested in higher scores. Compared to the general population those people who are high scorers—that is, those who are in the top interval of the trend—are less happy with available snack products, more happy with available alcoholic drinks. On the other hand, those who are in the top tertile on Family Autonomy express a large product deficit on alcoholic drinks and less on snack products. Thus, the table shows that wanting different types of snacks which are not available is highly and positively correlated with Rejection of Authority, Family Autonomy and Informality: Whereas wanting different types of alcoholic drinks is highly and positively correlated with Family Autonomy, Liberal Sex, Elegance in the Home, Rejection of Authority and Equal Opportunities for Women. All of these trends have evolved quickly in the late 1960s and the 1970s.

Table 10.2 Product deficit indexed by social trends.

	Alcoholic drinks	Snack products
Informality	74	157
Rejection of Authority	144	152
Equal Opportunities for Women	136	82
Family Autonomy	178	122
Desire for Elegance in the Home	192	97
Liberal Sex	164	89

The link between wanting different snack products and Informality gives us our greatest opportunity in new product development. Informality has increased substantially in the last 10 years. This trend relates to the growing acceptance of disorder in our everyday lives, and rejection of formal hierarchies and formal modes of communication. Table 10.3 shows an example from the UK of the increase of this trend over the last decade. Informal structures are developing in everyday life. The family offers one of the most striking examples of this. In most families 30 or 40 years ago, the father was the master in the house and

Table 10.3 Indexed increase in Informality trend since 1974.

Year	Informality
1973	100
1977	108
1980	117
1982	183

was authoritarian. Nowadays, for a substantial proportion of the population, the family is viewed and experienced more as a cell within which its members interact. The trend towards informality is correlated with Family Autonomy and Rejection of Authority.

The world of fashion offers another example of informal organization. Fashion has changed so profoundly and so swiftly precisely because of its informality, with few formal structures capable of standing in the way of change.

In our everyday lives, informality is more and more obvious. People are contesting central authority, they are asking for more and more information, they are moving towards everyday behaviour which is more flexible; less importance is attached to ritual. The importance for marketing is particularly obvious when one is dealing with population segments far removed from formal order and hierarchic authority. Leading brands and corporations are well advised to avoid giving the impression of using their power in a way that is contrary to the interest of consumers. Statements such as 'everybody is doing it', 'the best', 'the biggest' are less effective these days. It is becoming more productive to engage in a dialogue which appears to take consumer needs into account.

There is tremendous scope for new products, given this greater degree of informality. Not only do those expressing the need for greater informality say they are not satisfied with existing snack products, they are above average consumers of snack products (see Table 10.4). Some 36 per cent of those high on Informality say they also try new products as they come on the market.

Table 10.4 Eating of take-away foods by high scorers on Informality.

	Total %	High on Informality %
Eat Chinese take-away	37	45
Eat Indian take-away	23	28
TOTAL respondents	1533	329

Informality, spontaneity and the importance of new family structures are probably also linked to the growing preference for drinking at home rather than in public houses (see Table 10.5). Certainly, drinking wine is frequently linked with Informality as well as many other social trends (see Table 10.6). We must

Table 10.5 Prefer to drink at home or with friends than in a pub or club.

	1975 %	1982 %
Strongly agree/agree	31	42
Disagree/strongly disagree	41	39
TOTAL respondents	1518	2171

note also that women in particular have moved towards preferring to drink at home rather than in a pub (50 per cent of women *v.* 34 per cent of men) and more women than men drink white wine (69 per cent *v.* 61 per cent). These findings lead us to hypothesize an opportunity for more wine-type products marketed for women.

Table 10.6 Consumption of white wine by high scorers on specific trends.

		High scorers			
	Total	Informality	Individuality	Equal Opportunities for women	Family Autonomy
Drink white wine frequently (%)	11	15	16	16	17
TOTAL respondents	1525	329	414	584	422

USING SOCIAL CHANGE FOR PRECISE TARGETING FOR NEW PRODUCTS

So far we have seen a link between social trends and the desire for new snack products and a link between social trends and the possible retargeting of white wine. Rather than deal with all 35 social trends we turn now to the simpler classification of consumers.

From our study on social attitudes we have been able to group people into broad categories according to their values and beliefs. Some companies are familiar with these MONITOR Social Value Groups and make use of them in their new product innovation strategy. There are seven groups in all, each with distinct attitudes (and, as it happens, differing behaviour patterns as well). Two groups epitomize our understanding of the new, inner-directed consumer and the traditional outer-directed consumer, namely Self-explorers and Belongers.

We see a typical Self-explorer as someone living a 'progressive' life within the mainstream of society; he/she is probably well educated and in a professional occupation, living as an equal with whomever partner he/she chooses. Our evidence suggests that the number of Self-explorers in the population is increasing. As a consumer, Self-explorers demand high-quality produce (and are prepared to pay for it). Such goods may have to be durable, natural, healthy, individualistic, creative, pleasurable and not wasteful in energy terms.

In contrast, the Belonger is the traditional consumer in the 'Happy Families' mode. A Belonger would be living a conventional life in which the husband has a career and the wife looks after the home (perhaps working part-time for some extra cash). Belongers look for satisfaction from their home and family and are concerned to provide for their needs and be seen to do so. They comprise the largest proportion of consumers, although their overall numbers appear to be in slow decline. As consumers, Belongers are highly brand-conscious and brand-loyal. They are conservative in what they would consider buying and would not try anything new and different until it was well-established among their friends.

Two such different groups of consumers require different marketing approaches. The message to new product innovators is that they must first understand the needs and motivations of their consumers and then target their products accordingly. Out of the many aspects of the marketing function let us focus on three as illustrations.

Purchase Criteria

People with predominantly inner directed values, such as Self-explorers, tend to be far more discerning consumers than those with outer-directed or sustenance values. We have already mentioned that Self-explorers look for inherent quality rather than respond to product image. They are concerned about buying products where the ingredients are listed; they look for foods which state nutritional value; they buy goods which are more expensive but last longer. They consume more fresh food than other consumers and are at the forefront of the 'simple food' movement (albeit paying more for it).

In contrast to the inner-directed consumer stands the more conventional, outer-directed consumer, like the Belonger. Belongers want products that are well known, low in risk, nothing too 'way out', acceptable amongst their peers. More adventurous outer-directed consumers would also look for products incorporating new forms of technology.

The primary purchase criterion of sustenance-directed consumers is that of price, although with their tendency to shop locally on a day-to-day basis they pay more for their products overall.

Branding

The growth in acceptance of own-label products has been led by inner-directed consumers, who look for quality irrespective of the name on the front of the package.

The established big brands continue to appeal to the outer-directed consumer looking for status and conformity. They also retain the loyalty of the sustenance-directed consumers who tend to trust large, well-known companies.

Advertising

Traditionally, advertisements have appealed to the basic outer-directed or sustenance values—to feelings of prestige or belonging. Such advertising policies may be appropriate for consumers who come into these categories, but they only serve to alienate the inner-directed consumer. Of all of the groups, the inner-directed consumers such as the Self-explorers are exposed to the widest range of advertising media, yet they are the most disenchanted with conventional advertising. Inner-directed consumers are concerned about honesty and openness in advertising and appeals to status or quality based on a manufacturer's name alone would have no meaning for them. They look for information about the products and services they buy and rely on their own understanding and intuition when purchasing. Any manufacturer or retailer hoping to attract this group has to present itself as being straightforward, caring, open about its products and, above all, not manipulative.

SOCIAL VALUE GROUP ANALYSIS AND FOOD CONSUMPTION

By adding only 15 questions to any survey we are able to categorize respondents into Social Value Groups. The Social Value Groups themselves were developed in 1979 via factor and cluster analysis, the same exercises having been conducted on previous and subsequent survey data and the results found to be comparable.

We have recently added these 15 Social Value Group questions to the questionnaire we put to the Family Food Panel, which has been operated by Taylor Nelson since 1974 (see Appendix II for technical details). Before showing the link between these Social Value Groups and actual consumption of food in the home, it is necessary to look at some recent changes in eating patterns. For example, the level of total meal occasions in the home has increased since 1977, with this increase coming about entirely from total snacks. Across the same time span, total formal meals have remained relatively static.

Since 1981 there has been a further substantial year-on-year increase in snacks. Breakfast, lunch and tea have remained the same while there are stronger signs for evening meals.

The pattern of snack eating is not consistent across all snacks, with declines or marginal growth for some products. The real increase has come about from crisps, nuts and savoury snacks, chocolate-coated biscuits, countline biscuits, but above all fresh fruit.

In contrast, if we look more generally at some of the markets which have decreased, we find that there are many which may be described as traditional foods. One that is typical within the dessert area is custard, which has dropped quite severely. Against this, yoghurts and fresh cream desserts have increased in consumption. Both are illustrative of goods whose increased or decreased usage can be linked to the Social Value Groups.

Comparing again the individualistic, inner-directed Self-explorer with the traditional, outer-directed Belonger, we find that there are some major differences in food consumption: Self-explorers have an above average consumption of foods in 'progressive' and often growing markets, such as pasta and yoghurt. Belongers have an above average consumption of foods in more traditional and sometimes declining markets, such as custard. Hence we have been able to single out innovative, inner-directed consumers, such as Self-explorers, as the leading figures in the desire for and acceptance of new, exciting food products.

CONCLUDING REMARKS

I have attempted to show the rationale behind using social change analysis as a framework for new product development.

My examples are drawn from observations of product deficits—where consumer needs have not been fulfilled—and of the link between certain social trends and social values and existing consumption. I have demonstrated a new type of targeting in order to isolate opportunities in food and drink.

My concluding remarks relate to the organization of new product development. I believe that in future the marketing function will have a more strategic role, and will take a central position within firms, with far more input into and influence on the other functions within the organization—including new product development. The time horizon for the marketing function will be extended to include the time taken for research, product development, testing and production. I can imagine that even with fast-moving consumer goods, the time horizon could be at least 10 years.

Marketing companies in the future will have to become more robust, resilient and flexible. We visualize that there will be newer approaches to new product development. Smaller groups and cybernetic command structures will replace rigid hierarchical structures and inflexible company policy. Small project teams including people from all groups—research and development, planning, marketing, advertising, public affairs and sales—might help in the process of new product innovation. New product concepts will have to reflect constant switches and changes in consumer attitudes.

If marketing people can respond to social change then new products and services will emerge in response to the considerable reserves of unsatisfied needs among consumers. All obstacles to innovation will have to be tackled; new solutions will have to be sought where new social needs, new technology and new economic facts of life meet. I believe it will be the marketing function which will look for these solutions.

APPENDIX I: TAYLOR NELSON MONITOR

The aim of MONITOR is to help companies anticipate the social movements that may affect their environment in order to minimize marketing threats and maximize opportunities.

The Taylor Nelson MONITOR service is responsible for an annual survey which covers a broad range of topics in order to assess the implications of past, present and future sociocultural change. The survey has been conducted yearly since 1973 amongst adults in England, Scotland and Wales. There are now more than 17 000 interviews in the databank. Much of the information covered in the survey is continuous, which allows for quantitive comparisons of attitudes over time. For example, MONITOR has a wealth of data collected over the years on shopping, media habits, health and nutrition. All of this helps MONITOR to assess the direction in which consumer behaviour is changing. MONITOR also detects any newly emerging attitudes or forms of behaviour that should be incorporated into the survey from continuous qualitative research.

The basis of the philosophy of MONITOR is that the determinants of a society's characteristics are the individual's values and concerns.

People's different attitudes and values give rise to differences in lifestyles which in turn set up different needs and means of satisfying those needs. MONITOR surveys allow us to measure and track those current trends which, in time, will manifest themselves in social changes. In this way, we are able to help manufacturers and public agencies to distinguish between short-term fads and long-term shifts in consumer behaviour.

Social trends are the primary tool used by MONITOR in examining how British society is changing. Social Value Groups are also valuable in this context, but their main use lies in the area of market segmentation. In addition to categorizing survey respondents by standard demographic variables, MONITOR also allocates them to discrete groups based on shared attitudes and values. This form of classification has often proved to be more discriminating than systems based on demographic data alone. By looking at the consumers within a particular market by their Social Value Group types, it is possible to infer their attitudes and behaviour in relation to everything from their concerns about advertising to their frequency of shopping.

Social trend and Social Value Group data are interpreted by MONITOR within a marketing context to provide companies with a means of understanding their consumers and the society in which they live.

APPENDIX II: TAYLOR NELSON'S FAMILY FOOD PANEL

Taylor Nelson's Family Food Panel was the first continuous research programme to be developed for providing information on the usage of different foods and drinks inside the home. It was established in March 1974 and so trends and changes in eating and drinking patterns within the home can be traced since then. It is, therefore, possible to relate housewives' behaviour when preparing and serving meals today with that in previous years.

The Family Food Panel consists of a sample of 2100 households, representative of all private households in Great Britain. Each household reports, for

2 weeks in every 6 months, upon all foods and drinks consumed within that household by all family members and any visitors during that fortnight. The Family Food Panel provides a massive databank on usage of foods and drinks within the home. It provides information on what is being served at mealtimes, who is eating which foods throughout the day and how the food is prepared. It records all food consumed at home; whether prepared by the housewife or somebody else, or is brought into the household.

The Food Consumer
Edited by C. Ritson, L. Gofton and J. McKenzie

CHAPTER 11

Food Processing

RAY BRALSFORD

INTRODUCTION

One of the striking factors about modern food products is the extent to which they have been 'processed'. However, food processing has been the preoccupation of man ever since he progressed beyond the point where he consumed his food at the place where, and at the moment when, he acquired it. In other words, one of the main purposes of food processing—to preserve food and regulate its availability between times of harvest—is as old as civilization itself. Only the techniques have changed; field-dried pulses, for example, have largely been replaced by the quick-frozen variety (with a consequent improvement in nutritional value).

However, a modern purpose of food processing is to provide the food with new attributes which appeal to the consumer and which extend beyond that of mere year-round availability.

The first of these new attributes is convenience. The traditional ordered society, with the housewife spending long hours in the kitchen, preparing meals to be consumed by the whole family, has gone. At the same time, the taste for more exotic food has grown. The food manufacturer has therefore tried to provide products which are attractive but easy to prepare, often in single rather than family quantities, and often in portable eat-on-the-move form which is more appropriate to the modern lifestyle.

The second new purpose of food processing arises from the consumer's increasing preoccupation with the role which food plays in living a healthy life. Products which claim, for example, 'extra protein', 'added vitamins' or 'low calorie' rub shoulders with those which claim absence of additives ('natural') and, at the very extreme, with the crankier foods sold in some health food stores.

The last main change in the function of food processing has been to use new knowledge in food science and technology to provide foods which are cheaper alternatives to traditional foods, either by making use of new raw materials or by processing traditional raw materials in new ways. Margarine as an alternative to butter, and the up-grading of carcase meat by a variety of techniques, are examples.

This chapter is therefore concerned with the way in which the properties of foodstuffs, in the scientific sense, when combined with the properties which the modern consumer finds appetizing, have lead to new technology and hence to foods in new forms.

APPROACHES TO FOOD PRESERVATION

After harvesting, most foods, whether animal or vegetable in origin, begin to deteriorate. They lose nutritional value and eventually become inedible. Some, such as grain or root crops, if stored under suitable conditions, deteriorate very slowly. Others, such as meat, unless treated in some way (i.e. processed), will deteriorate very rapidly. The consequent limitation of choice of foods at some times of year led in the past both to tedium and nutritional imbalance, particularly from the lack of fresh vegetables and the scarcity of meat in winter.

Ancient man's successful attempts to extend the shelflife of perishable foods were presumably based largely on accident, but the principles embodied in drying, smoking, pickling, salting and curing are largely unchanged in modern processing.

Two main processes are involved in the deterioration of food after harvesting. The first is the chemical breakdown of the food constituents, largely under the influence of enzymes which catalyse the breakdown process. For example, protein molecules, which are generally very large, are gradually broken into smaller fragments and eventually into the basic amino-acids of which they are composed. Fats similarly may break down into fatty acids and will react with oxygen in the air to produce rancid off-flavours.

The second process of deterioration involves the microorganisms—bacteria, yeasts, moulds—which are present in all foods. They may use the food for their own growth and multiplication, and in the process change its composition. Sometimes the effect is benign, such as the conversion of sugar by yeasts into alcohol or the acidification of milk by bacteria in yoghurt making. More often, the effect of microorganisms is undesirable and may at best make the food unpalatable, or at worst lead to illness or death if the food is ingested. This may arise either by the continuing multiplication of microorganisms in the human body or by the toxic effect of the chemical by-products of the microbiological growth.

There are two processing approaches to the problem of microbiological deterioration of foodstuffs. The traditional approach is to change conditions in the food so that, even though they are still present, the harmful microorganisms are unable to grow and multiply. This is the basis of traditional salting, drying, pickling, etc. The same principles are still used today but our improved scientific understanding enables us to use the techniques in more subtle ways.

The second approach to food preservation is to leave its composition more or less unchanged but to kill all the microorganisms, usually by heating them, and subsequently to store the food in a way which prevents ingress of new organisms from the environment. This is almost exclusively a modern technology, manifested, for example, in canned foods and UHT milk.

Occasionally, some new and traditional processes make use of a combination of the two techniques and heat is used to kill the relatively benign yeasts and moulds which cause food spoilage, whereas the composition is adjusted to prevent growth of the potentially pathogenic bacteria which may survive the heat process or may contaminate the food subsequently. Preserves are a traditional example of the use of the dual techniques; yeast and moulds which may enter the jar after the jam has been made will grow on the top of the jam, but the much more harmful pathogenic microorganisms will not.

Avoidance of chemical deterioration of foods during storage is generally more difficult than the prevention of microbiological spoilage. Canned, dried or even frozen products which may be perfectly safe to eat after long storage will eventually be rejected because taste or texture has become unacceptable. The measures open to the food processor are few. He can remove the air from the pack by evacuating it, or by replacing it with another gas such as carbon-dioxide so that those reactions involving oxygen are slowed or stopped. He can chill or freeze the product so that the chemical processes are slowed down. He can remove water so that those chemical processes which require it are inhibited, but this may be an unsuccessful strategy because the removal of water may promote other chemical or physical changes in the food. He can blanch the food, i.e. subject it to a relatively mild heat treatment, so as to reduce enzyme activity and hence slow down the chemical processes.

MICROBIOLOGICALLY STABLE FOODS

As already discussed, one of the simplest methods of food preservation is to remove the water. Without water, microorganisms cannot grow. Field-dried pulses and grains depend in part for their stability on their relatively low moisture content. Sun-dried meat, fish and fruit were probably amongst the earliest of the true products of this process. However, in the modern context, dehydration suffers from a number of disadvantages.

First, the dried food itself is generally inedible without rehydration. Second, the process of rehydration is usually long and is not completely reversible, to the extent that the original texture is never regained. Finally, the process of removing water generally tends to remove flavour at the same time.

Field-dried peas suffer from all the disadvantages mentioned and yet command a small but enduring market. Soaking the dried peas overnight, with the rehydration process aided by the addition of a soda tablet, results in a product which is not remotely similar to the fresh variety but one which is still prized by many. An early improvement offered to the consumer, which at the same time helped keep the cannery occupied in winter when fresh vegetables for canning were in short supply, was to rehydrate the peas, thereby destroying their stability, and then to sterilize them in cans. These are the so-called 'processed peas', a term which distinguishes them from field-fresh, canned 'garden peas'.

A later significant development in dehydration was the process known as freeze drying. The technique had long been used in laboratories to remove water from biological material, whilst at the same time avoiding shrinkage and the loss of structure which accompanies air drying. In the mid-1960s the opportunity was seen for the large-scale commercial application of the process to foods. The advantages foreseen for the process were as follows:

(1) Dehydration saves the cost of shipping large quantities of water during distribution.
(2) Because the water is frozen before drying in a vacuum chamber, every part of it evaporates from the position it originally occupied in the food. It does not migrate in the liquid state to the surface of the food, as it does in simple air drying, carrying with it dissolved components which are left on the surface, slowing both the dehydration process and subsequent rehydration. Large food pieces could therefore be dried, which would nevertheless rehydrate quickly.
(3) The low temperature at which the process is carried out would yield better flavoured products.

In reality, freeze drying has found much more limited application than was once foreseen. The process turns out to be very slow in relation to the capital cost of the necessary plant. This is particularly true for large food pieces where the advantages of the process were expected to be greatest. The main application of freeze drying has therefore been for products which are granular in form, such as freeze dried coffee and other high-value products such as prawns. Furthermore, although freeze drying produces solid products which rehydrate quickly, the essential irreversibility of the process remains. The water associated with proteins cannot be replaced once the protein has been 'denatured' by dehydration and the original texture is never regained.

Another problem associated with freeze drying is that all dehydrated products, even good ones, are very resistant to attempts to move them up-market. Blind tests have proved that careful (and uneconomic) processing can produce dehydrated, canned and frozen versions of the same food which are indistinguishable. However, if tasters are aware of the process used, they invariably rank dehydrated as lowest and frozen as highest in quality.

For both quality and cost reasons, factory dehydration remains, therefore, a process which is mainly limited to products which either start as liquids (milk, coffee) or can be presented in a finely divided form (potato powder). For liquids, an effective and cheaper process than freeze drying is spray drying. The liquid is first concentrated by partial evaporation of the water and is then sprayed into the top of a large tower where it meets a current of hot air. The droplets lose moisture quickly and because of this high rate of evaporation stay relatively cool, in spite of the high air temperature. The dried droplets are collected as a powder from the base of the tower.

Many spray-dried powders are difficult to reconstitute because water will not readily 'wet' the product. This problem is overcome by 'instantizing', the main component of this process being the agglomeration of the fine particles into larger but loosely adhering aggregates which disperse more readily in water. In the case of milk powder, this may be achieved by sending the powder for a second time through the spray dryer with a smaller amount of water which re-wets the particles sufficiently to stick them together. This water is finally removed by further drying.

In the case of dried potato powder, dehydration of the cooked and mashed potato takes place in a stream of hot air. The problem of the dried soluble materials left behind at the food surface leading to slow drying rates would be particularly troublesome in the case of a sticky product such as potato. In this case the solution is to share the liquid with a large amount of already-dried powder. This so-called 'add-back' manufacturing process takes some time to establish, and once running it is clearly difficult to change operating conditions on the plant.

Dehydration, in which nearly all of the water is removed from the food, is not the only way in which water can be made unavailable for microbiological growth. Water can be diluted, in effect, by dissolving other materials in it. As well as the well-known effects of lowering the vapour pressure and depressing the freezing point, the presence of dissolved materials makes the water more difficult to get at by microorganisms. The 'water activity' is said to have been reduced, to a point between the value of unity which it has in the pure form, and zero which is the value it would have if the solution were so concentrated that there was virtually no water present at all.

The reduction of water activity is greater, the greater the number of molecules which are dissolved in the water. Consequently, pound for pound, small molecules like salt are more effective than large molecules like sugar. However, not all microorganisms react in the same way when water activity is reduced. The most responsive are the bacteria, and a reduction to 0.94, equivalent to a 50 per cent sugar solution or a 10 per cent salt solution, is sufficient to stop growth of all the most serious food-poisoning bacteria such as salmonella. At a water activity of 0.75, most moulds will cease to grow and at a value of 0.65 only a few 'osmophilic' yeasts will continue to survive.

The next step in rendering a foodstuff microbiologically stable is to make it more acid. At pH 4.5, bacteria will not grow, but yeasts and moulds will. Traditional fruit preserves depend for their effectiveness on a combination of water activity reduction, acidity and some heating in order to kill yeasts and moulds. Jams, if processed properly, will only grow moulds or ferment once they have been opened and reinfected from the air. However, in some of the newer, reduced sugar jams now on sale, there is insufficient water activity reduction for stability and in this case the law permits the use of a preservative. Sorbic acid is very effective for this purpose; it is the natural preservative occurring in some fruits.

Understanding of the role of water activity, acidity and preservatives, combined with pasteurization to reduce initial microbial loads, has improved greatly in recent years. Chill distribution slows down the rate of microbiological spoilage still further and this has led food technologists to try to design products with the appearance and appeal of a fresh product, but with an extended shelflife of a few weeks. Some chill cabinet desserts such as yoghurts are based on these principles, but success in the broader food area has been limited. Usually, the compromise between a recipe which is tasty and one which is stable is uneasy; there is a limit to how much salt, sugars and acidity can be incorporated into a recipe. Furthermore, the 'dilution' of water with other, more expensive ingredients does not benefit costs. The concept of 'intermediate-moisture' foods, half way in moisture content between normal foods and the fully dehydrated form, has found success only with food for cats and dogs, and that to a limited degree.

The most commercially successful of the modern methods of food preservation has been the use of deep-frozen distribution. This, too, can be seen as a water activity reduction technique, where the liquid water necessary for microbial growth is largely replaced by ice, and the low temperature greatly slows all the process of deterioration. (Chill distribution provides the benefit of slower chemical and microbiological changes but water activity is not affected.)

There are very few practical limitations to the types of foodstuffs which can be successfully frozen, except for some fruit and vegetables with high water content where the delicate structure tends to be destroyed. Recipe products are limited in quality only by the cost which the market will bear. Furthermore, rapid harvesting and freezing result in products which are superior in nutritional value to many of the unprocessed vegetables which have made their slow progress to urban outlets through the older ambient temperature distribution system.

To summarize, the techniques for extending the life of foodstuffs by controlling the moisture, water activity, acidity and temperature have changed very little. Attempts to extend their application into new areas have met with limited success, because of the compromise between stability and quality. Deep-freezing, although not new in principle, has expanded rapidly because it does not suffer from the same limitations.

STERILIZED FOODS

The alternative strategy for the prolongation of the shelflife of perishable foods is to accept their basic instability but to exclude, from the food's environment, the microorganisms responsible for the deterioration. The technology necessary for this technique is not traditional and has developed more recently than the much older compositional techniques.

The simplest method for killing microorganisms is to heat them and this has the advantage that, at the same time, many of the enzymes responsible for chemical spoilage are also inactivated. Heating has the disadvantage that it unavoidably cooks the food to a degree which may be undesirable.

The processing must also be done in such a way that reinfection from microorganisms in the atmosphere is impossible. Heating in a hermetically sealed container is the simplest method, but this only became a process of feasible, safe application after the invention of the tin can. The related kitchen process of 'bottling' is only safe where the acidity and/or water activity of the contents, such as fruit, are low enough to prevent the growth of surviving or adventitious pathogenic bacteria.

The key to good canning practice is known in the trade as the 'botulinum cook'. The most hazardous of all the pathogens in the heat-preserved foods is *Clostridium botulinum*. This belongs to a class of organisms which are widely present in the environment. They can also occur in a dormant spore form which is very heat-resistant compared with the alternative vegetative growth form. If spores of *Clostridium botulinum* should survive the heat sterilization process and begin to grow, a toxin is produced which is one of the most potent known to man, but which is tasteless and odourless. Consequently, all sterilized products, which do not have a composition which would prevent the growth of *Clostridium botulinum*, receive a time/temperature treatment which guarantees its destruction. Many products are overcooked as a consequence.

It is to the credit of the UK canning industry that no case of botulism has ever been ascribed to canned products produced in this country, though there have been deaths attributed to imported canned products. Occasionally, things can go wrong in a cannery, and cans may miss being processed altogether or imperfect seals can lead to reinfection. Such accidents invariably lead to a whole collection of microorganisms being present and the can contents will be noticeably 'off' and are unlikely to be consumed—an inbuilt safeguard, though not good for a manufacturer's reputation!

Developments in heat sterilization processes have been directed in recent years towards overcoming the problem that the 'botulinum cook' exceeds the degree of cooking which would lead to the best quality for many products. Obviously, this problem is greater the larger the container, since the 'botulinum cook' must relate to that part of the contents, near to the centre of the container, which heats most slowly. Two approaches to the problem are possible: change the shape of the container, making it flatter so that heat penetration is more rapid; or carry out the heating in a more suitably shaped vessel and transfer the food aseptically to the previously sterilized final container. Both approaches are in use.

Shorter processing times are one of the benefits claimed for sterilizable plastic pouches for foods. It was claimed by the originators that, because of the pouch's flatter shape, heat treatment could be reduced and quality improved. The higher the quality of the starting ingredients the truer the statement turns out to be. For example, a range of gourmet dishes are available which are much in demand on expeditions to the remoter regions of the Earth. In the West, however, the sterilizable plastic pouch does not seem to be a serious threat to the deep-frozen equivalent. This is not the situation in Japan, where the lack of any well-developed frozen distribution system has enabled the sterilizable pouch to become an important part of the processed food market.

Another reason for the relative slow growth of pouches is that packaging costs are higher than those for frozen products, and production rates on the relatively sophisticated handling and processing plant tend to be low. Pouches have not, therefore, provided the means of attacking frozen products on the basis of the cost of the frozen distribution system.

Half-way between the traditional can and the flexible sterilizable pouch are the so-called semi-rigids, consisting of plastic-lined aluminium containers pressed to a variety of shapes. Again, the costs are high and commercial applications have been limited.

The alternative approach to reducing the overcooking of sterilized products is aseptic processing. One of its earlier and more bizarre manifestations involved the use of a work area, the whole of which was maintained at a pressure equivalent to a domestic pressure cooker. Products were sterilized in equipment designed to give fast heating at temperatures well above normal boiling point, but equivalent to those in conventional canning. The products were then filled into cans which were sealed without further cooling by workers dressed in diving-suit-style clothing. This system has not been widely adopted!

In France some soups are sterilized by moderate heat treatment and then filled aseptically into conventional cans. As well as giving a product which more closely matches the French taste, there are clear energy savings for the manufacturer. In the UK, the steady growth in alternative preservation techniques, notably freezing, has meant that the canning industry has been in slow decline for many years, and the consequent competition resulting in an inadequate profit margin has meant that little money has been available for investment in new canning plant, even where operational and quality benefits are shown. In any case, unlike the French, the UK consumer's norm for good soup has become the overcooked variety, made necessary by the requirement of the 'botulinum cook' and the limitations of can geometry in the conventional process. However, in the UK aseptically canned ready-made custard is the exception because the in-can sterilization process would result in a product so caramelized that it would be unacceptable as custard.

Once the process of sterilization has been moved outside the container, not only can the total heat treatment be reduced, but there is then freedom to choose the combination of time and temperature which best suits the product. In other words, a given degree of sterilization can be achieved either by heating at a relatively low temperature for a relatively long time, or by heating at a much higher temperature for a much shorter time. For solid foods this benefit is largely of theoretical interest only, because the geometry of the food pieces themselves impose a new limit to the rate at which they can be heated; solids are also difficult to fill aseptically into the final container. For liquids, however, the situation is different.

It is generally true that sterilization at higher temperature for a shorter time produces less flavour change. Milk and milk-based products are particularly sensitive to 'boiled milk' flavours when sterilized, and this fact led to the development of the ultra-high-temperature-short-time sterilization process—UHTST, later abbreviated still further to UHT. Here the milk is heated very quickly by passing it as a thin film through a special heat exchanger; it is kept hot for a few seconds, and then quickly cooled before filling aseptically into its final pre-sterilized containers, usually cartons but occasionally plastic bottles. Alternatively, the milk may be heated by injecting steam into it, holding it at a high temperature for a short time, and then by spraying it into a vacuum chamber where the evaporation of water from the milk droplets quickly cools it again.

Processed under the best conditions, UHT milk is a very acceptable alternative to the fresh product and much superior in flavour to products produced by the older in-bottle sterilization process. Unfortunately, it is not always the best starting material which is subjected to the UHT process. The primary concern of most dairies is with milk for fresh consumption, where it is vital that the inevitable microbiological growth, which eventually turns the milk sour, has been limited as far as possible during the period between the milk leaving the cow and its leaving the processing plant. Any batch of milk which may be less than ideal in this respect tends to be directed towards the UHT plant, resulting in a product which is perfectly satisfactory from the microbiological point of view, but which has less than optimal flavour.

UHT processing is one area where techniques can be foreseen which will improve the quality of such products still further. But to the extent that food processing is an attempt to produce long shelflife versions of what were formally short shelflife products, it is difficult to see many new directions. The frozen-food manufacturers have already covered most of the ground; the manufacturers of dehydrated and sterilized products have improved their quality but what they offer is more an acknowledgement of the limitations of their own distribution systems than a positive consumer benefit.

Non-thermal sterilization, using high-energy radiation from a radioactive source, is practicable; but the technique is more likely to find application in reducing losses due to infestation of raw materials stored in the tropics than it is in new consumer foodstuffs. The emphasis in food process development seems likely to be much more in the direction of improved quality and convenience and much of this development is likely to be centred on the chill cabinet. The concentration of the grocery trade in a relatively few national multiples with their central warehousing systems offers the opportunity to shorten the supply chain. Prepared meals and complete recipe dishes, with only a few days' shelflife at chill temperatures, provide the consumer with quality and true convenience.

CONVENIENCE FOODS

Freezing peas may be a way of extending their availability beyond the 6 weeks or so that they can be obtained from the greengrocer. The justification for freezing potato chips has to be different.

We have largely passed beyond the time when housewives felt guilty about using convenience products; when it was said that she needed to be allowed to add an egg to the packet cake mix in order to be able to live with her conscience. (The real reason that she was allowed to do this was that manufacturers had great difficulty in producing a low-cost dried egg which would work in a cake.) But just as important as reduced kitchen labour is the modern need for products which can be quickly prepared on an individual rather than a family scale, or can be eaten informally whilst travelling, working or relaxing.

The rapid change in the confectionery market towards eat-from-one-hand chocolate bars and away from bags of sweets is one manifestation of this change in lifestyle. An even more significant change is the very rapid growth in the market for savoury snacks; these, too, are beginning to evolve from bags to one-handed format.

Much of the development in savoury snacks owes its origin to the development of a processing technique in an industry quite separate from food manufacturing, namely the use of extruders in the conversion of basic plastic raw materials into the finished product. An extruder consists essentially of a barrel containing one or more rotating screws. The material to be extruded is fed in at one end of the barrel; it is mixed and compressed as it is conveyed to the other end of the machine by the screw, where eventually it emerges through a shaping die. Extruders have been widely used in the food industry for over a century for applications as diverse as stuffing sausages or shaping pasta. But these were relatively unsophisticated machines, operating at low temperatures and pressures and incapable of working at all outside a limited range of recipes.

As a result of improvements in their design by the plastics industry, extruders eventually found their way back into food applications. It now became possible to extrude cereals, particularly maize, at high temperature and pressures so that they expanded at the die face, producing new kinds of textures and shapes. Such products were marketed as breakfast cereals or savoury snacks.

In the past few years cook-extruders have developed to the point where many of the mixing, heating, cooling and concentration operations, which would previously have required separate pieces of processing plant, can take place within a single machine, offering significant advantages in capital and operating costs. Crispbreads are now made almost exclusively by cook-extruder techniques. Recipes for savoury snacks are governed much more by what tastes good and much less by what can be persuaded to go through the extruder.

A more portable and more convenient snack than earlier products is the long-life savoury sausage, the market for which is growing steadily. This product uses a combination of the preservation techniques described earlier, combining water activity management with the use of preservatives and heat processing in flexible pouches.

One of the prizes which has eluded snack manufacturers so far is the attractive combination in a single product of the crispy texture of fried potato or cereal with the chewy textures of cheese or meat. This last property only occurs in foods which contain significant amounts of water and the problem of preventing the migration of moisture from the chewy to the crispy parts of the product has so far defied solution.

Another problem, which *has* been solved, is how to provide a hot snack away from the kitchen environment. Here, the quality limitations of dehydrated ingredients become a virtue, and the addition of hot water to a plastic pot containing dried noodles, vegetables and other ingredients provides at the same time both the means of rehydration and the means of heating a shelf-stable product in a convenient portable form. The great success of this type of product shows that it fulfils the need for a hot snack in a way which is difficult to improve on for simplicity.

Back in the kitchen, one of the most widely used groups of convenience products is the beverages. One of them, cocoa, has remained as stubbornly inconvenient a drink as it was when first marketed in the last century, but it is available under a different legally prescribed name as Drinking Chocolate. This is an instantized version of cocoa in which the fine particles are agglomerated with sugar, by processes similar to those used for milk powder, in order to make them disperse more easily in hot milk.

Coffee powder, obtained by drying coffee liquor by any of the processes described earlier, is probably the most widely used convenience beverage. Modern techniques for retaining the volatile flavour components make the product an acceptable alternative to the traditional beverage for most people in the UK. In contrast, no generally acceptable instant tea has been produced. In part, this must be due to the long tea-drinking tradition which sets high standards of acceptance for the convenience product. But more important is an essential difference in the chemistry of the two beverages. The flavour and aroma components of coffee are reasonably stable in the infusion (freshly infused coffee can be kept warm for quite a long time without much loss of quality) whereas the flavour components of freshly brewed tea continue to interact, even in the cup, and the drink quickly loses its freshness. For the instant beverage manufacturer, coffee therefore presents less of a problem. There is time to complete the process of concentration and drying before too much has changed in the chemistry, whereas a tea infusion has spoiled long before the drying stage has been reached.

A second problem of instant tea manufacture arises from the fact that the dissolved solids content of tea is very much lower than that of coffee and it is difficult to produce a product which, using normal kitchen techniques, can be measured in the small quantities needed for a single cup. Convenience in tea has therefore followed the alternative route of packaging in tea bags, which is perhaps a less satisfactory solution to the problem, but one which is sufficiently attractive to consumers to cause them to buy more of their tea in bags now than they do in the traditional packets.

PROCESSED FOOD AND HEALTH

People are increasingly concerned about the role which food plays in the search for longevity and a healthy life. Television, books and newspapers all recommend consumers to adopt a wide variety of dietary changes. Some of this advice is based on scientifically reputable work, but much more of it is presented by individuals whose credentials are doubtful and whose motives are mixed. The outcome seems to be that people have become greatly confused about what action they should take, if any, if they wish to improve their diet.

One of the main confusions is the role which processed foods play in determining whether a diet is healthy or otherwise. Many commentators use the word 'processed' as though it were synonymous with bad nutrition, even though there is no case to support this view. There is no difference in concept or in reality between factory processing and kitchen cooking; both involve the mixing, heating and concentration of food ingredients. There is, of course, bad food processing just as there are bad cooks. In both cases the quality of the final outcome depends not only on technique, but also on the quality of the starting materials. The food processor has added value to his product by undertaking some of the work of preparation on behalf of the customer and it is the customer who decides whether to trade off convenience against quality, or to pay the price which provides them both.

It is also implied by their critics that food manufacturers make much profit out of selling inferior products to a gullible public. On average food manufacturers achieve levels of trading profit, i.e. profit before interest payments and tax, which are less than 5 per cent of the sales value of their products. Food retailers do significantly better than this.

The confusion of thought is illustrated by the story of a young home economist who presented an excellent dissertation extolling the virtues of tofu, the traditional Japanese food prepared from the curds of soya milk. She believed that this would be a better way of using soya protein in the Western diet than some of the new food technology based uses. Apart from the fact that most of her taste panellists liked tofu as little as they liked textured soya protein, the striking thing about the 'natural' use of soya was that the description of its method of preparation was exactly the kind of description which would be given to a food process engineer who had been asked to build a factory plant — except that the process engineer would have used hygienic stainless steel vessels, whereas tofu seems for preference to be produced in unhygienic wooden tubs.

As far as processing itself is concerned, therefore, the health implications are not different from the health implications of cooking. The food-processing industry certainly poisons fewer people than do incompetent cooks, and the best processes, involving detailed specification and control, destroy less of the nutritional properties of their raw materials. In some cases the processor restores the lost nutritional value by, for example, adding back vitamin C lost during processing of instant potato powder; the vitamin C lost from overboiled, badly stored potatoes is not replaced.

The second area of health concern about processed foods is the use of materials which are loosely called additives.

Whatever their origin, all foodstuffs consist of a mixture of chemicals, the main components of the mixture usually being fat, protein and carbohydrate. The minor components include vitamins, minerals, colours, emulsifiers and antioxidants. All of these classes of minor components occur naturally in foods, but when they are added by the food processor they are called 'additives', irrespective of whether the added substance is chemically the same (as it is in many cases) or different from any naturally occurring substance.

There are three main reasons why the food manufacturer uses these ingredients:

(1) Some materials are used as aids to processing, e.g. the oil used to release tablet jellies from their moulds, or the anti-foaming agent used to prevent pineapple juice frothing during processing. The presence of such materials in the end product is incidental and is not designed to enhance the product.

(2) Some additives perform specific technical functions in the product. For example, an emulsifier is necessary in order to produce a stable emulsion such as in butter or margarine. In the first example, the emulsifier is present naturally; in the second case the emulsifier, which may be equally natural in origin, has to be added.

Antioxidants prevent oxidation and rancidity of the oil in fried snacks.

In order to ensure a proper set, it is sometimes necessary to add extra pectin to jam because some fruits do not contain enough. The level of acidity also affects setting properties and this is adjusted by the use of citric acid.

(3) The ingredient may be used to enhance the appeal or performance of the end product, e.g. colours, flavourings, or anti-caking agents in powders such as salt.

In the UK what may be added to foods and used in their preparation is covered by extensive and detailed regulations. These are made jointly by the appropriate ministries, but notably the Ministry of Agriculture, Fisheries and Food, and the Secretary of State for Social Services, acting within the framework of the Food Act 1984 and on the recommendations of expert advisory committees. Similar regulations apply in other Western countries. These regulations are based either on long traditional use or on extensive safety testing supported by convincing evidence of technical need.

The subject of additives nevertheless raises intense emotion. Undoubtedly, some additives cause allergic and other physiological reactions in some people. But so do unprocessed foods. The sufferer from coeliac disease cannot tolerate the gluten in wheat flour; many people react to the lactose in milk; strawberries cause rashes in others. Moreover, chemical analysis of some traditional foodstuffs shows that they contain substances which would be classed as poisons or carcinogens if they were present at higher concentrations. But most traditional foodstuffs have not benefited from detailed analysis or long-term controlled safety testing. We are aware, therefore, only of the most acute results of eating some natural foods which we avoid because we class them as poisonous. We know nothing about the less immediate results of eating particular unprocessed foods over many years.

On balance, logic says that we ought to feel safer with additives, which have been extensively tested and the use of which is properly controlled, than we do with the millions of other naturally occurring chemicals in foods of which we know nothing. The world consists entirely of chemicals and they cannot be divided into 'natural' and 'unnatural'.

Regrettably, however, because food manufacturers will supply consumers with the products they demand, a great deal of attention is being devoted to marketing products to which the claim 'natural' can be attached which in reality have no significance in terms of health and nutrition. To this extent, therefore, effort which could have been devoted towards fulfilling more meaningful dietary changes is being wasted.

The important safeguard for consumers is that labelling should be understandable and informative. Food manufacturers as a whole have no objection to improvements and changes in food labelling legislation but they do require that such labelling should be simple, appropriate and fair. This was one of the aims behind the use of the 'E' number, which is intended to convey to those consumers who are interested precisely what a particular additive is and also to convey that the additive has been subject to exhaustive evaluation by the regulatory authorities of the European Community. In contrast, 'E' numbers have become synonymous with undesirable food constituents and attention has been drawn away from more important dietary areas.

An increasing number of manufactured products claim positive health attributes. The commonest of these are the slimming products. There are two ways to make a low-calorie food. The first is to reformulate it so that it contains less fat, less protein and more carbohydrates (calorific values 9.0, 5.2 and 4.8 cal/g respectively).

Margarine, like butter, is composed of 82 per cent fat and 18 per cent water. Lower-calorie margarine has half the calories of normal margarine or butter and is composed of 41 per cent fat and 59 per cent water. This achievement requires subtle processing of the fat and the use of special stabilizers and emulsifiers. Its composition explains why its use on hot toast is less than satisfactory.

Ice-cream appears on many slimmers' lists because of the large volume of air it contains. Low-calorie sugar is also available in which the sugar granules have been expanded so as to contain half the amount of sugar and the lost sweetening power has been replaced by a non-sugar sweetener; the product can be used in exactly the same way as sugar, but half the calories are consumed.

The arguments surrounding the role of fats in the high incidence of heart attacks in the West are complex and are covered in some detail in chapter 3. There is, however, general agreement that the disease involves more factors than diet, e.g. smoking and exercise, but many people believe that there is strong evidence that reducing fat intake and, where possible, substituting polyunsaturated for saturated fats can significantly lower blood cholesterol levels and the attendant risk of developing heart disease.

Whatever their origin, all oils and fats have closely similar chemical composition. Every molecule of fat consists of three fatty acids connected at one end by a bridge of glycerol. Each fatty acid consists essentially of a chain of carbon atoms, usually twelve to twenty-two in number, and each carbon atom may have up to two hydrogen atoms attached to it (the very last carbon atom in the chain can have three hydrogen atoms attached). If all the carbon atoms have their full complement of hydrogen atoms attached, the fatty acid is said to be saturated; if at more than two points along the chain there are hydrogen atoms missing, the fatty acid is said to be polyunsaturated. The only difference between naturally occurring fats from different sources lies in the length and the degree of saturation of the fatty acids which make up the various triglycerides of which they are composed. As a very broad generalization, fats of vegetable origin tend to be less saturated than fats of animal origin, although there are notable exceptions to this rule.

Cows can be persuaded to some small degree, through their diets, to produce milk fat which is less saturated than normal. The margarine manufacturer has the advantage of choosing fats from a wide variety of sources which he can blend, and by careful control of the way he cools and crystallizes the fat he can produce a product which, although fairly solid in appearance, contains a surprising amount of liquid polyunsaturated oil. He can further extend his choice of raw materials, and hence his ability to make products with very specific properties, by processes which rearrange the order of the fatty acids in the triglycerides (a controlled and speeded up version of a process which occurs in the kitchen chip pan). He can separate the triglycerides into high and low melting fractions and he can replace part of the missing hydrogen in the unsaturated fats.

NEW SOURCES OF FOODSTUFFS

Ever since the prediction by Malthus that population growth must eventually outstrip his ability to feed himself, man has dreamed not only of improving agricultural output, but also of finding entirely new sources of food which might be less susceptible to drought and storm. The agriculturists have had great success in improving crop yields, and by that success have reduced the value of even the very modest achievements of scientists seeking food from unconventional sources.

Much of the search in the past 20 years for new raw materials, whether based on novel farming techniques or on new processing technology, has rested on the assumption that there is a world shortage of protein. More recently the realization has dawned that man's nutritional need for protein is not as great as was once thought and that, if he can be supplied with sufficient calories, he will almost inevitably consume sufficient protein. The development of local agriculture is the only practical long-term method of achieving this result.

Work on proteins in recent years has therefore shifted in emphasis towards reducing its cost, which is much greater than that of carbohydrates or fats in the Western diet. The new proteins are essentially of two distinct types. First, there are those proteins from sources which have not previously formed a significant part of the human diet; examples are soya protein and microbial protein. Second, there are proteins from traditional sources which might be used in novel ways; examples are wheat gluten or milk protein, both of which might be used to simulate meat.

The best known of the new protein sources is soya. After the Second World War huge acreages in the USA were transferred from maize production to soya production, partly for the value of soya oil. The high-quality protein by-product left after oil extraction is still used mainly as cattle food, but in the 1950s two methods were discovered which were designed to enhance the value of soya protein by giving it the properties of meat. The simplest of the two is the process in which de-fatted soya flour is transformed, in a cook-extruder of the type described earlier, into a fluid mass under conditions of high temperature and pressure. The mass expands and sets as the pressure and temperature fall at the extruder die face to form textured vegetable protein. The second, more complex, process involves dissolving the protein in alkali and then 'spinning' it into an acid medium to produce fibres by processes similar to those used for man-made textile fibres. These fibres can then be binded together by further protein, e.g. egg white, in order to mimic meat.

Attempts to market products in which soya meat replacements are the major component have been largely unsuccessful. Textured vegetable protein is not seen by most consumers as an acceptable alternative to meat and the more sophisticated spun product suffers from high costs. However, soya protein is increasingly used as a minor ingredient in many recipe products where it can be used to perform specific technical functions, such as binding water, imparting texture or stabilizing emulsions.

Projects which aimed to extract protein from the leaves of a variety of plants which would be otherwise useless for human nutrition have not resulted in products which are attractive as food ingredients unless expensive purification processes are used.

The second group of unconventional proteins are produced by techniques which fall into the fashionable category described as biotechnology. The oil companies have devoted much effort to the development of processes in which protein-rich microorganisms are grown on substances which are derived from oil feedstock. The microorganisms are then harvested and the protein extracted for further processing. One large plant is operating commercially in the UK producing microbial protein for animal feed. However, the process is almost certainly not economic and the value of the project probably lies in the knowledge being gained about the operation of very large scale fermenters for applications in more profitable non-food areas.

An alternative microbial process, developed by a large UK food company, starts from food-grade material as substrate for the growth of microorganisms. The aims have been to produce protein which would be more readily acceptable as human food than one derived from an unconventional feedstock and, moreover, whose physical form would be ideally suited to mimic meat. The organism eventually selected was a *Fusarium*, which is a filamentous organism found in soil. The substrate can be starches from wheat, potato, etc., or sugar from molasses. The resulting protein, because of its filamentous nature, can be converted into an excellent substitute for meat. This material has undergone extensive safety testing and has been accepted by MAFF for use, at the moment, at an agreed restricted level in the food chain.

The future for biotechnological processes as a major new source of food raw materials is unclear. The new processes, if they are to compete with the traditional raw materials, seem to face quite serious cost problems. The process energy content of the new proteins is high and this component of total cost is rising more rapidly than the general rate of inflation. In contrast, prices of conventional agricultural raw materials are generally increasing less rapidly than inflation. Moreover, the USA and Canadian surpluses in maize and soya have lead to a great increase in world trade in these commodities, and their relative cheapness has enlarged their use in animal feeds, thereby slowing down the rate of inflation in meat prices, and reducing the attractiveness of the new meat substitutes.

The immediate application of biotechnology in the food and drink industry is likely to continue to be mainly in the production of alcoholic beverages, and in the subtle modification of food ingredients, such as the use of enzymes to modify cheese flavours, or to alter the sweetness of glucose syrup.

SUMMARY

Although processed foods are increasing as a proportion of total food consumption, most of the process techniques used have not changed very much in principle from those used traditionally in kitchen cookery. However, the ability to select and control the properties of food ingredients is growing rapidly.

New processed products will be bought more for their quality than for their long shelflife. Consumers will increasingly buy processed foods for greater convenience and less as reserve supplies. The food processor can supply the higher quality demanded once he is released from the constraint of paying for part of his processing costs out of savings on raw materials. Where the consumer is looking for other attributes, such as low calories, or portability, quality may be compromised.

Dramatically new sources of raw materials are not likely to emerge from the new technologies. Improved quality and performance will.

The Food Consumer
Edited by C. Ritson, L. Gofton and J. McKenzie

CHAPTER 12

Catering—Food Service Outside the Home

GEORGE GLEW

Since the Second World War, eating outside the home has become part of the way of life in many industrialized countries and catering has become one of our largest service industries and a major employer of labour. In the UK the catering industry is the fourth largest employer of labour, with more than 2 million workers (nearly 10 per cent of the workforce). An increasing proportion of consumers' food expenditure is on eating out. In the mid 1970s consumers' expenditure on eating out in France was 20 per cent of total food expenditure, in Sweden 15 per cent and in the UK 12 per cent (Glew, 1980). In the USA it is expected that by 1990 nearly half consumers' expenditure on food will be on meals eaten outside the home. In all industrialized countries the catering industry will play an increasingly important part in meeting the consumers' needs for food in the last fifth of the twentieth century.

In this chapter, consideration will be given to the reasons for the growth of this industry, the variety of catering outlets now available and how the industry tries to meet the consumers' needs.

CATERING AS A SERVICE

Catering involves the provision of a service. The term 'foodservice' is used instead of 'catering' in North America, and graphically describes what the industry is about.

Catering developed as part of a social class structure in which elements of the community were employed to provide a cooked food service for those in higher positions in society. The higher social classes in all countries did, and still do, purchase in some way a catering service, and they do not spend time in the cooking and serving of food for themselves. The cook has always held a position of importance in any group or household and there are clear indications of the social importance of groups of likeminded people eating together and sharing food cooked for them. In modern households one member of the family cooks for the others and the psychological link between the provider of food and the consumer is strong. A service is being provided for others to enjoy and there is a large difference between the purchase of a manufactured product and a service. Social interaction between the provider of the service and the consumer of the service takes place; there is little or no similar direct interaction between the manufacturer of a product and the consumer or user of the product.

The consumer of food in a catering establishment is not only concerned with satisfying physiological needs—important psychological needs must also be satisfied. The service provided for the consumer takes place in a situation which is unique and unrepeatable; the environment, state of hunger or satiety, mood, state of health and many other factors for both provider and receiver of the service are unique to one occasion in a lifetime. Hence the provision and receipt of service involves extremely complex interactions and the physical object around which the service revolves—the meal—can be of far less importance than is often realized. The meal can become the vehicle for expression of feelings and views totally unrelated to the occasion. A 'meal' becomes an 'experience', the least important part of which may be the physical nature of the food itself. Food in hospitals may be criticized, not because it is intrinsically poor, but because the consumer has other worries which find expression in criticism of the food. In other instances, food which may clearly be intrinsically poor (e.g. many 'meals-on-wheels' supplied for the elderly) is rarely criticized by the recipient because the physical entity of the meal itself is the least important aspect of the occasion; the most important is the social contact with the provider of the service.

THE BIRTH OF GASTRONOMY

In the last hundred years the hotel and catering industry has become specialized in the same way as every other industry. Increasing specialization has resulted in the plethora of eating-out establishments that now exist, from snack bars and take-away burger outlets to *haute cuisine*, family restaurants and the many ethnic restaurants that now exist. The provision of residential accommodation has become specialized and often separated from the provision of food. It is now normal in the USA to hire overnight accommodation in a building (hotel/motel) which has no foodservice. Obtaining food is not difficult, but the food is provided quite separately by another organization, usually down the street, as a separate specialized enterprise.

This development has been evolving since the modern age of catering started in the 1870s. The increasing complexity of life coming in the wake of the industrial revolution was reflected in the increasing demands of consumers for complexity in meals. Revolution in manufacturing techniques, together with changes in the class system and society in general, also produced a revolution in the way in which hotel and catering services were provided. The modern hotel originated at the beginning of the nineteenth century to cater for the travelling nobility who moved about with a large entourage of servants. The *hôtel garni* consisted of rented apartments in which the personal servants created a home-from-home and maintained a supply of foods to meet the head of the household's wishes. Taylor and Bush (1974) have described the emergence of spas and resorts and the provision of elegant accommodation which made the nobility feel at home. It became inconvenient to travel with a large entourage, and personal servants began to be replaced with a staff who performed the same functions but who were employed by the hotel. Rich but not noble customers began to appear who wanted to copy the lifestyle of the nobility and so this principle has slowly pervaded the whole of society.

The 'mine host' of the inn and tavern of former centuries had now to specialize in order to meet customer demands. This point is well illustrated by the work of César Ritz and Auguste Escoffier. Ritz was the hotelier *par excellence* who formed a partnership with Escoffier, the chef who revolutionized the system for meal production. Ritz and Escoffier worked at the end of the last century and during the first two decades of the twentieth century (Page and Kingsford, 1971). Ritz worked throughout Europe, and in a relatively short working life (he died at the age of 51) changed the ethos and mechanics of management in the hospitality industry. He worked at the Savoy, Carlton, Claridges and Hyde Park hotels in London, the Grand in Rome, the Kaiserhöf and Augusta Victoria Baths in Wiesbaden, the Frankfurterhöf, the Grand Hôtel des Thermes, and the Grand Hôtel Palermo and its restaurant in Biarritz. At the same time, Escoffier was developing a new system for coping with the demands of the increasingly sophisticated restaurant customers in some of these and similar hotels. He claimed to have developed the 'partie system' in which kitchen activities were divided into single groups of product, such as soups, meats, fish, sauces, pastry and so on. Each group was the responsibility of a single chef skilled usually only in the production of items of that type. The components of a meal were therefore produced separately at different 'stations' in the kitchen and were assembled under the eye of a single, highly skilled chef. By using such a system it became possible to produce very complex meals of high quality with staff of limited ability.

Escoffier, writing in 1907, drew attention to 'the gradual but unquestionable revolution' in catering which was taking place which he attributed to 'the great impetus given to travelling'. 'In regard to the traditions of the festal board, it is but 20 years since the ancestral English customs began to make way before the newer methods .... But these new-fangled habits had to be met by newer methods of cookery—better adapted to the particular environment in which they were practiced'. He outlined the new habits as 'modern society ... partaking of light suppers after the theatres ... the well-to-do flock to the [restaurants] on Sundays in order to give their servants the required weekly rest'. And 'restaurants allow observing and of being observed, since they are eminently adapted to the exhibiting of magnificent dresses'. He explained that 'scarcely one old-fashioned method of cookery has escaped the necessary new moulding required by modern demands'.

The partie system was designed to cope with a clientele demanding new dishes, new gastronomic experiences, as part of the exciting new technological world developing around them. Escoffier bewailed, 'What feats of ingenuity have we not been forced to perform, at times, in order to meet our customers' wishes?' In the pursuit of newness and excitement his consumers placed demands on him that made him exclaim, 'But novelty is the universal cry—novelty by hook or by crook!' (Escoffier, 1907).

In a relatively few years, eating out was revolutionized in the sense that the new systems designed for food production allowed a large increase in the variety and subtlety of meals to be achieved with limited staff resources. However, it must be remembered that these delights were the privilege of the *haute monde* and that the vast mass of the population merely looked on.

POPULAR CATERING

The Lyons family had entered the business in the 1890s with a philosophy which allowed anyone to partake of their form of hospitality, provided that they could pay for it. One of the first 'popular' hotels in London was the Strand Palace, opened by Lyons just prior to the First World War, and it made a profit. The accommodation included breakfast, and no tips were expected. The Lyons Teashops and Corner Houses, the Kardomah chain and others in the 1920s and 1930s gave the middle classes, and even the working classes, the opportunity to enjoy the pleasures of eating out at a cost which they considered bearable.

A widening of the catchment population for catering and hotels continued with store restaurants opening to provide a midday meal for shoppers, and the growing habit of families taking an annual holiday. This was frequently at a boarding house or pension and, in Britain, invariably at the seaside. The seaside resorts such as Brighton developed because of an interest in the health-providing properties of seawater, but were transformed into leisure centres also providing food (often in take-away form, such as fish and chips) and drink for the masses. If family finances did not allow a hotel or boarding house stay in such a resort, then day trips to the coast were possible from almost any town in Britain.

Increasing disposable income and higher aspirations continued after the Second World War with part of the outlet for these aspirations being in eating out. The crucial facilitator of international tourism was, of course, cheap air travel. For the first time it became possible to travel long distances relatively cheaply and enjoy the food of different cultures. In addition, restaurateurs became mobile in large numbers and moved from country to country seeking their fortune. Today it would be difficult to think of a town of any importance in Northern Europe which did not have a Chinese restaurant. Indian restaurants are very popular, as well as the more localized Italian, French and Spanish cooking traditions. The consumption of ethnic foods by large sections of the population, particularly the young, is a recent development and has even been adapted to the take-away trade so that such foods can now be consumed at home, though cooked by specialists elsewhere.

FAST FOOD

In the fast food area of catering, the Americans have had a major influence. The fast food revolution has rolled in waves over Europe from North America. The idea that fast food is new is a clever marketing exercise to increase its attractiveness. Fast food has been with us since the Romans and it is difficult to imagine anything faster or more convenient to eat than fish and chips in Britain, frankfurters in Germany and Denmark, tacos and tortillas in Mexico, and crêpes in France. However, the difference between these traditional fast foods and the modern concept is the method of marketing and selling these products, and this development is wholly American.

The hamburger, or beefburger, was first used in the USA as a means of using cuts of meat which were unsuitable for roasting or frying. The meat was chopped into small pieces, a process which broke up the connective tissue, then pressed together to form a meat patty. These products were easy to prepare and cook and were relatively cheap.

The fast food product has been taken over by the food-manufacturing industry and is produced on a large scale to exacting specifications and with the application of the full panoply of modern quality control methods to ensure achievement of predetermined quality standards. Some of the largest chains of fast food operations are owned by food manufacturers: Wimpy in the UK by United Biscuits Ltd; in the USA, Burger King by Pillsbury, and Pizza Hut by Pepsi Cola. The large-scale production of burgers, pizzas and other such products is a manufacturing operation in which art and craft play no part. The skills of the food scientist and technologist, microbiologist and engineer are paramount. At the point of sale to the consumer, the product is cooked freshly in simple apparatus, the operation of which can be taught in a few minutes. The product for final sale is assembled by unskilled workers from prepared materials. A variety of prepared dishes always accompany the food—cake, milk shakes and so on. The attributes needed of the workers are speed and the ability to concentrate, to smile at customers and to keep the product moving.

Fast food operations are retail operations—they are shops selling a small range of cooked dishes. High Street locations are sought, and garish shop fittings to impart the required razzmatazz are part of the exercise. Good packing is vital—

heat-insulated boxes, hand and face cleaning tissues, ketchups and condiments are part of the product.

The consumer's image of fast food also includes the use of modern materials in design, bright colours and, above all, cleanliness. All surfaces are easy to clean; rubbish is out of sight; uniforms are crisp; hair covering is obligatory for staff; bright lights and music give the impression of pulsating modernity. The vendors of fast food are not only selling food but excitement and experience too. The food stereotyping is disguised by the ambience. Fast food restaurants are places in which to meet friends, make a date, satisfy hunger, be kept warm and entertained—fast food is not intended to provide a gastronomic experience for the cultivated palate.

CATERING IN THE WORKPLACE AND AS A PART OF SOCIAL SERVICE

Meals have to be provided at the workplace in factory and office canteens, in schools and other educational establishments, in hospitals and prisons, in the armed forces, in homes for aged persons and other places where there is some difficulty in consumers providing their own meals. Sometimes consumers are unable to provide their own meal because they are too far from home, particularly in the middle of a working day, to travel home to consume the meal. In industrialized societies factories are frequently isolated from the areas of towns where people live. Similarly, in many countries the school day spans a meal period at midday and the pupils are too far from home for it to be possible for them to eat their midday meal there. In hospitals and other institutions food has to be provided for both patients and staff, frequently on a 24-hour and 7-day-a-week basis. In many countries elderly and disabled people are able to receive midday meals in their own homes by means of a meals-on-wheels service; in this case the consumer is too frail or disabled to prepare his/her own meal and the catering industry is therefore involved in providing a particular service.

In all cases mentioned above the provision of a meal is an adjunct to some other activity; people do not go to school, hospital or factory to be fed. The main purpose is to learn, to be made well or to work and the consumption of the meal is an ancillary activity which is essential to prevent the onset of hunger.

The problems of food provision in these circumstances are very different from the problems associated with food provision for consumers eating out in the commercial sector for pleasure, or in the course of business. One of the main differences is frequently in the numbers of people to be fed in one place and over a very short time period. It is often the case that many hundreds, or even thousands, of meals have to be served and eaten within a very short time span, hence the use of the term 'mass feeding'. The equipment, staff training and financial arrangements for catering in these circumstances are often very different from those which prevail in the commercial sector, where the success of a catering operation can be judged on the basis of profit in relation to financial turnover. In many countries where there is massive state support for hospitals, schools and in feeding the elderly, simple financial criteria as a measure of success are not available. The real cost of providing a hospital meal is frequently difficult to calculate when the operation is run by the state. The value of the meal to the consumer is almost impossible to measure, hence there are no easy yardsticks to use in assessing the success or failure of such an operation.

In the next sections, catering in a non-commercial environment will be discussed and some of the non-financial advantages to the consumer will be outlined.

Catering in the Workplace

The beginnings of catering in the workplace can be found in the factories of the early Industrial Revolution in Britain where provision was made for workers to cook their own food (Curtis-Bennett, 1949). During the nineteenth century a number of philanthropic, Quaker factory owners, such as Cadbury and Rowntree, provided workers with canteen facilities, but it was not until during the First World War that the provision of a hot meal for factory workers became common in many countries. This provision increased even further during the Second World War and since that time many countries have acquired legislation which requires factory owners to provide meals for their workers (if they employ above a certain number).

It would be satisfying to believe that there was a close relationship between the provision of food at work and working efficiency. However, such a direct relationship is very difficult to prove conclusively. Much will depend on the nutritional status of individuals and what they consume outside the factory canteen. Buzina *et al.* (1972), El Batawi (1972) and the FAO (1976) have all published work which would indicate that higher productivity and lower absenteeism could result from the provision of food at work. Other writers have claimed that fewer industrial accidents and better productivity are associated with the consumption of an adequate breakfast prior to work (Tuttle and Herbert, 1960; Brooke *et al.*, 1973). It is now becoming clear, however, that generalizations about the effects of different feeding régimes on populations as a whole are often unsatisfactory. The point at issue is the nutritional status of an individual within a population. The knowledge that the population of a country as a whole is adequately nourished does not mean that there may not be large numbers of individuals within that population who are undernourished for a whole variety of reasons. Such reasons may include poverty and the inability to buy sufficient food, ignorance of the right types of food to purchase, or likes and dislikes for different foodstuffs. Nevertheless, it might be safe to conclude that workers in a country where the population as a whole is undernourished are likely to benefit from the provision of meals at their workplace, leading to reductions in illness, absenteeism and labour turnover, together with increased productivity.

School Feeding

Similar arguments to those used in the previous section might be applied to the feeding programmes associated with schoolchildren. Such programmes have been described as mainly 'substitute' feeding programmes (in that they replace part of the day's food intake) by the World Health Organization (WHO, 1975). In industrialized countries this attitude to school feeding might be adopted as a generalization, but in the developing countries the meal at school may play a very important part in the growth and wellbeing of the child. In former times, in both Britain and the USA, school meals services were established to help allay the effects of severe poverty amongst the working population. The Destitute Children's Dinner Society was founded in 1864 in London and had opened 58 dining rooms by 1869. In New York, the Children's Aid Society is recorded as serving a meal at lunchtime as early as 1853, and the Chicago Board of Education started their meals service in 1910 (Van Egmond, 1974). In developed countries, the reasons for school meal provision have changed, particularly in the last 20 years, with more emphasis being placed on meal provision necessitated by distance from pupils' homes and the increasing proportion of the female population in fulltime employment.

Few countries have declared nutritional targets for their school meals but those which do usually aim to provide about one-third of the daily nutritional requirements in the school meal. The meal at school is usually taken at midday; however, in the USA increasing emphasis has been placed in recent years on the school breakfast programme, the argument being that the children who arrive at school hungry are more disruptive and less likely to benefit from the education being provided than children whose appetites have been satisfied. There is no doubt that even within the richest countries there are pockets of poverty and ignorance, where a need for supplementary feeding of children at school would appear to have considerable importance.

As with industrial catering, it is difficult positively to identify measurable advantages to be obtained from the establishment of the school feeding programme. Even in India, the World Food Programme (1978) has been unable to identify improved scholastic ability or academic achievement as a result of school feeding programmes. However, in both Japan and India claims have been made that the physical measurements of children participating in school lunch programmes are greater than those of children not provided with school lunch (Roy and Rath, 1972; Kato, 1977). If malnutrition is a problem in any area, then it is likely that the school lunch will have some significant advantage to the children involved. However, when children are well nourished the provision of food at school may well have no other influence than to satisfy the immediate physical and psychological effects of hunger. In Kenya, Pieters (1974) (using anthropometric and biochemical measurements) found, in an area where malnutrition was not a problem, no significant differences between the two halves of a group of schoolchildren, one half of whom were participating in a school lunch programme.

Hospital Catering

The person who made nursing into a profession was also the person who introduced and organized the first foodservice system in hospitals. During the Crimean War (1854-1856), Florence Nightingale, invited Alexis Soyer, the chef of the Reform Club in London, to visit her hospital at Scutari and organize the provision and cooking of food for the patients. Her patients were not only suffering from their wounds, but also from a very inadequate diet. Soyer laid the foundations of the hospital catering industry which has developed alongside other improvements in health care.

The patient in hospital may need to consume food with certain nutritional properties as part of the treatment being received. A number of diseases, for example diabetes, can be controlled by a combination of medical and dietetic treatment. Supplying the needs of the consumer in hospital can have a major and immediate effect on the health of the consumer. Perhaps because of the clear relationship between nutritionally inadequate food and disease, the food in hospitals has been subjected to more careful study from a nutritional viewpoint than in any other area of the catering industry. It is not only to patients suffering from specific nutritionally related diseases that the nutritional content of food is of great importance; nutritious food is also essential to those patients who are chronically ill and likely to be in hospital for months or even years. In 1963, Platt *et al.* reported on the quality of the food in 153 hospitals in England and Wales. They found that the best food and service was provided in the best small hospitals, and that, in general, the larger the hospital the lower was the quality of the food served. There was a tendency to overcook vegetables, and service methods for keeping food hot for long periods resulted in nearly complete loss of vitamin C from potatoes and 75 per cent loss from green vegetables. In hospitals for chronically sick patients, the three main meals contributed only about one-third of the British daily recommended allowance of vitamin C (30 mg per day). The researchers believed that these defects were caused by faults in catering administration, problems of communication and of distribution and inadequate training of medical and nursing staff which tended to concentrate on the nutritional requirements of specific diseases and neglected the general requirements of patients. In many hospitals new catering methods which have been introduced in the last decade have helped to alleviate some of these problems. More rapid cooking of vegetables in high-pressure steamers helps to retain the vitamin C and more rapid distribution systems when food is plated in the kitchen are among changes which have taken place.

Feeding the Elderly

Provision for the care of the elderly is at present largely of concern only in industrialized countries, where the proportion of the population which lives to the age of 80 or over is still increasing. Part of the package of care for this group must be the provision of food. Elderly people are either looked after by their own family, live alone or with their ageing spouse, or become resident in a home for elderly people. The problems of food provision for those people being cared for in a home are not dissimilar from the problems of feeding the chronic sick, in that all their nutritional requirements must be in the food supplies provided by the home.

Most of the elderly are well able to care for themselves in their own homes, even though social isolation becomes an increasing problem, but between about 10 and 15 per cent, because of physical or mental impairment, require support from outside their home by voluntary or state organizations. In many countries a service of food provision for elderly people in their homes has been developed which has become known as 'meals-on-wheels'. In these services, hot meals are delivered to the elderly person's home with a frequency which varies from one to seven meals per week. It has been stated (Exton-Smith *et al.*, 1972) that if fewer than four meals per week are delivered their contribution to the total nutritional intake is not significant. A major problem relates to the nutritional effect on the food of keeping it hot during delivery, and a study in Britain (P.A. Management Consultants, 1973) showed that only 27 per cent of the meals were delivered in under 2½ hours. Many types of delivery container are used, from simple insulated boxes to electrically or charcoal heated containers. No standards for nutritional content of the meals exist and there is good evidence that heat-labile vitamins are lost from these meals, mainly at the prime cooking stage but also during transport to the old people's homes (Davies *et al.*, 1973; Armstrong *et al.*, 1980).

The provision of meals-on-wheels presents one of the most difficult catering problems because, although the meals required are large in number, and could be produced on a large scale, the consumers are widely dispersed geographically throughout the community. Some of the methods that can be used to overcome these problems will be discussed in the next section.

CATERING AND THE CONSUMER

Near the beginning of this chapter, some ideas in relation to catering as a service were explored. The close physical as well as psychological relationship between provider and consumer of the service was contrasted with the much larger gap between the consumer and the manufacturer of goods. However, the caterer is a manufacturer as well as a provider of a service. He makes a dish from simple materials and his skill adds to the value of the materials. This concept of 'added value' is one which the consumer finds difficult to grasp, so that people criticize the catering industry for 'overcharging' because everyone knows the cost of the raw materials and many expect the finished product to cost very little more. This is an attitude which is unique to catering; such complaints are not raised against the cost of a refrigerator or dishwasher, where the cost of raw materials is a very small fraction of the purchase price and the value added by the manufacturer is considerable. However, a little thought makes it clear that a variety of overhead and recurrent costs such as rent, taxes, energy, staff, as well a profit margin, must increase the purchase price of a meal to perhaps several times the cost of the raw materials.

For full satisfaction the consumer must, after consideration of the price paid, be happy with the intrinsic quality of the meal and the service and ambience within which it is provided, and he will take these into account when making a judgement on whether he has received value for money. A further consideration relates to the purpose of the meal. Is the meal for refuelling or is it part of a leisure period? Speed of service and ease of eating may be important in a factory canteen or in the midday break of a city office worker, but the same worker will display a very different attitude in the evening when eating out with family or friends.

The caterer, in trying to satisfy the consumer's expectations, is forced by rising labour costs to look to technology to allow him to increase the proportion of the meal cost apportioned to labour at the expense of that portion which covers overheads and raw materials. If he is to keep the cost of the meal the same and increase his labour costs he must reduce his overheads and raw materials costs.

One partial solution to this problem is to separate food production from food service. Equally, catering is the last great industry to become industrialized—that is, to separate production from consumption. At present, even on a large scale, in school meals and other mass-feeding enterprises, the food is prepared, cooked and consumed within a few hours and usually under the same roof. All other industries have perceived advantages in separating their activities. In catering there are problems of keeping food hot between production and service (a process known as 'warmholding'). Warmholding often results in deterioration of flavour and appearance and also may reduce the nutritional value. In any event warmholding of cooked food can only be contemplated for a few hours.

Methods have been developed for divorcing food production from food service by cooling the cooked food. By reducing the temperature to just above freezing point (cook-chill catering) the gap between cooking and service can be widened to 5 days. If the food is frozen and stored at -18 C (cook-freeze catering) then the storage gap can be widened to weeks or even months. The creation of a gap between production and service eases the management and control problems affecting catering. The production of cooked and chilled, or cooked and frozen, foods can be controlled by the caterer himself or delegated to a food manufacturer. The main advantage is that the caterer is not trying to produce and serve food simultaneously. These systems based on chilled or frozen foods have been very successfully applied in large-scale catering, both in the public sector and in commercial catering (Glew, 1985a). In addition, there are many other forms of high-quality preserved food which caterers use extensively, such as canned and dehydrated products. Other types of preserved food will be used in the future as the development of the technology discussed in an earlier chapter is applied in the catering sector.

In addition to new food products and methods of presentation the caterer may also use new technology in the actual cooking processes. Microwave energy can be put to use by caterers, and other devices including continuous cookers and equipment controlled by microprocessors must be used to maintain standards and control costs.

Some consumers feel that food should be immune from change. There is a view that traditional is best and that caterers should use only traditional raw food cooked by traditional methods. This is impossible, even it if were desirable. Almost every variety of edible plant material has changed in genetic make-up in the last 30 years, so it is not possible to obtain the same raw materials as used by Escoffier and the great chefs of the past. Animals raised for meat are fed on different foods and reared in different conditions. The consumer must necessarily put his faith in the caterer and food technologist. Does it matter where or how the food was grown or how it was prepared and cooked if it meets the consumer's expectations when on the plate? Technology moves forward in every industry and the catering industry is no exception. No consumer would expect the caterer to cook all his food on a wood or coal fire, yet the change to gas and electricity as a form of heating was a revolution in its time. Caterers themselves are caught between the belief that the consumer is a traditionalist, yearning for what is safe and known, and the necessity to use technology to the full. Caterers tend to fear 'convenience foods' because the consumer may feel that he, the caterer, is taking short cuts. Part of the problem may be that the consumer of a pleasure/leisure meal is looking for a little magic surrounding the meal and the feeling that the kitchen staff are 'putting their all' into this 'special meal for you' is vitally important to them. The working day 'refuelling meal' does not have the same social significance so its method of production may be less important to the consumer.

THE CONSUMER OF THE FUTURE

It is likely that the proportion of meals eaten outside the home will continue to grow in the developed world. This trend will have an increasing influence on the agricultural industry, the food-manufacturing industry and the distribution and food-retailing business. In time these influences will be reflected in what the consumer can expect to receive. In addition, the consumer will influence what is provided by caterers. The consumer in the 1990s is likely to demand food from caterers which is more nutritious in the sense that it complies with the then current nutritional guidelines. All consumers are becoming aware that the food we eat can influence our health, and healthy eating out is likely to become increasingly important (Glew, 1985b).

Consumers are likely to demand fewer non-essential additives in food. Again, the realization that what we eat influences health will dominate the consumer's thinking. It is likely that increasing knowledge of the causes of allergic conditions will result in food scientists being able to remove such materials from food. These food products will be used by caterers.

Eating out will merge more with other forms of entertainment and leisure activity. An interesting development from the manufacturer of electronic games in the USA has resulted in the establishment of a chain of pizza restaurants for children where the eating is an accompaniment to a series of short performances by robot rabbits, bears and other creatures (Pizza Time Theatre). The performance is entirely computer-controlled, and it is expected that the production and service of the pizzas will also move in the direction of less human involvement.

Caterers are in the business of satisfying consumers wherever they may be and in whatever circumstances they find themselves. Eating out will undoubtedly continue and grow in a variety of forms and for many different reasons.

REFERENCES

Armstrong, J.F., O'Sullivan, K., and Turner, M. (1980) *The Housebound Elderly—Technical Innovations in Food Service*. Hotel and Catering Research Centre, The Polytechnic, Huddersfield.

Brooke, J.D., Toogood, S., Green, L.F., and Bagley, R. (1973) 'Dietary pattern of carbohydrate provision and accident incidence in foundrymen', *Proc. Nutr. Soc.*, **32**, 44a.

Buzina, R., Harvat, V., Broderac, A., and Vidacek, S. (1972) 'Nutritional status, working capacity and absenteeism in industrial workers'. In *Alimentation et Travail*, ed. G. Debry and B. Bleyer. Masson, Paris. 141-151.

Curtis-Bennett, N. (1949) The *Food of the People—The History of Industrial Feeding*. Faber & Faber, London.

Davies, L., Hastrop, K., and Bender, A.E. (1973) 'Ascorbic acid in meals-on-wheels', *Modern Geriatrics*, July/August, 390-394.

El Batawi, M.A. (1972) 'Food intake and work performance: a study on the effect of fasting on work output in the Eastern Mediterranean'. In *Alimentation et Travail*, ed. G. Debry and B. Bleyer. Masson, Paris. 261-265.

Escoffier, A. (1907) *A Guide to Modern Cookery*. 8th impression (1977). Heinemann, London. Preface.

Exton-Smith, A.N., Stanton, B.R., and Windsor, A.C.M. (1972) *Nutrition of Housebound Old People*. King Edwards Hospital Fund for London, London.

FAO (1976) *Feeding of Workers in Developing Countries.* FAO, Rome.
Glew, G. (1980) 'Background and trends in catering to 1990 in Western Europe'. In *Advances in Catering Technology*, ed. G. Glew. Applied Science Publishers, London. 3-15.
Glew, G. (1985a) 'Refrigeration and the catering industry', *Proc. Inst. Refrig. 1984-85. vol. 81.*
Glew, G. (1985b) 'Our industry 1979-84', In *Advances in Catering Technology*, three. Elsevier Applied Science Publishers, London.
Kato, M. (1977) 'Country reports on school food service, Japan', *International Workshop on Improving Nutrition and Nutrition Education Through School Food Service.* Agency for International Development, Washington, DC. 45-65.
Page, E.B., and Kingsford, P.W. (1971) *The Master Chefs.* Edward Arnold, London.
P.A. Management Consultants. (1973) 'Meals on wheels'. *Report to Department of Health and Social Security*, UK. 20.
Pieters, J. J. L. (1974) 'The effect of school lunches on nutritional state'. In *Alimentation et Travail*, ed. G. Debry and B. Bleyer. Masson, Paris. 27-35.
Platt, B.S., Eddy, T.P., and Pellett, P.L. (1963) *Food In Hospitals.* Oxford University Press, Oxford.
Roy, P., and Rath, R.N. (1972) *School Lunch in Orissa.* Council for Social Development, New Delhi.
Tuttle, W. W., and Herbert, E. (1960) 'Work capacity with no breakfast and a mid morning meal'. *J. Am. Diet. Assoc.*, **37**, 137-140.
Van Egmond, D. (1974) *School Foodservice.* AVI Publishing, Westport, Conn.
WHO (1975) 'Interorganisation meeting on expanded supplementary feeding programmes for vulnerable groups'. Nut./75.2, Geneva, 25-27 March. WHO, Geneva.
World Food Programme (1978) Committee on food aid policies and programmes', *A Survey of Studies of Food Aid.* WFP/CFA:5/5-C March. FAO, Rome.

The Food Consumer
Edited by C. Ritson, L. Gofton and J. McKenzie

CHAPTER 13

How People Choose Food: The Role of Advertising and Packaging

JUDIE LANNON

INTRODUCTION

The determinants of food choice are obviously complex and vary from product category to product category. While the most profound and significant influences on food choice are undoubtedly cultural and traditional, the presentation of food via advertising, packaging and other promotional activities under the control of the food manufacturer play a part. This chapter discusses the contribution of psychological and anthropological theory to the understanding of how mass communications affect choice. It also describes the range of different roles that advertising and packaging play in influencing people's choice of food.

Examples of how advertising was developed to meet particular marketing requirements are used by way of illustration.

THE NEED FOR A THEORY

First, however, we need a theory, or at least some guiding principles about how we feel advertising or, indeed, any form of persuasive communication works. Debates about the extent of the power of advertising take a number of different forms. On the one hand, the manufacturer of a packaged food product complains bitterly that an expenditure of several million pounds failed to sell his brand; *that* advertising conspicuously did not work. On the other hand, nutritionists and politicians of various affiliations complain that advertising creates consumer needs, teaches the public to eat what is nutritionally bad for them; *that* advertising is a force that manipulates gullible consumers.

That advertising has some effect on choice seems clear. However, exactly how the mechanism operates is rather less clear. What we do know is that advertising works differently in different product categories, on different people, in different marketing circumstances. There are no simple rules which are universally applicable.

Psychologists, sociologists, anthropologists and mass communications theorists have all had something to say about how advertising works. However, certain theories have been positively misleading while others have proved richly insightful; so it is worthwhile to have some discussion of the sources of current advertising practice.

Perhaps the simplest and most primitive model of how advertising works derives from classical learning theory which places considerable emphasis on the learning of facts, opinions and attitudes that subsequently modify behaviour. The central metaphor of this model is a hammer and nail, and an essential component is reinforcement, a repetition of the stimulus, hence the importance of repeating a slogan at regular intervals. Essentially all of these are 'transportation' models based on the assumption that advertising does things *to* people: that messages are carried as though on a bus to their destination where they will (once registered via recall or recognition measures as having arrived) have the desired effect.

'A Mars a day helps you work rest and play' and 'Have a break, have a Kit Kat' are two well-known and enduring slogans.

Another theory is based on what could be called a 'conversion' model, with the central metaphor based on the likelihood of regular 'Damascus road' experiences. Here the main emphasis is on convincing argumentation that assumes the power of rational persuasion. Nutrition arguments in advertising for manufactured goods have faltered occasionally because until recently people have simply not been interested; taste, variety or convenience were of greater interest or value. (The difficulties in conveying nutritional claims are dealt with later on.)

A third category of theories have in common the belief that, for a piece of communication to be effective, it must go through a series of stages: awareness of the message, belief in the message, and persuasiveness of the message as measured by research. While this model is likely to be appropriate for new technical innovations from new improved razor blades to new forms of computers, it does not explain why people continue to use products when they are unaware of the advertising, and fail to adopt products despite their proven awareness and comprehension of the advertising message.

And indeed, there is a common-sense validity to these models. So pervasive is advertising in modern society that one can understand the concerns of those who believe that its very presence in the media implies effectiveness. A regular finding in studies designed to assess the public's view of advertising shows a high degree of agreement with the statement 'Advertising makes people buy things they don't need or want' yet very little agreement with the statement 'Advertising makes *me* buy things I don't need or want'. Such paradoxical findings are not uncommon in attitude research: people like to think they are rational in their choices and yet assume that society is populated with gullible people (unlike themselves) who believe everything they hear.

But a considerable amount of experience suggests that advertising often works in other, more complex ways and that the 'common-sense' theories are limited in what they are capable of explaining.

The central theoretical flaw is the failure to account for the contribution of the receiver in the communication process. In this respect they are all 'passive voice' theories and share a perspective with sociological and religious theories, which position man as being helplessly acted upon. There is an assumption of a more or less passive receiver, a *tabula rasa* on which messages are printed. A more 'active voice' perspective (Douglas, 1982b) is more in line with current experience in mass communications studies.

In fact, the contributions of many other schools of psychology, including mass communications theorists, have produced considerable evidence that advertising does not work by the simple implanting of attitudes in the mind that lead to behaviour change. Cognitive theorists (Festinger, 1957) stress the resistance to dissonance. Since attitudes do not exist in isolation, the mind exerts a consistent tendency to maintain balance and resist change when confronted with conflicting information.

Probably the most widely observed instance of this is in the field of nutritional education. What people do *with* food information is a classic case of conflict. The aspects typically in conflict are, on the one hand, a sensual preference for the taste of the food in question, and information that says it is bad for one (sugar, chocolate, butter, etc.). Or the reverse may apply: a dislike for the taste of something conflicts with information saying that it is good for one. In both cases, the response, as predicted by psychological theory, is to ignore the incoming information because the imbalance is too great to accommodate.

Road safety, electricity conservation and alcoholic abuse campaigns, which all attempt to change behaviour on a mass scale, offer examples where information coming from the mass media is in conflict with existing beliefs or habit patterns, and it is far easier to reject or ignore the information than it is to change the habit.

Also of direct relevance is the work of Gestalt psychologists (Koffka, 1935) on perception. Individuals would not attempt, even if it were physically possible, to perceive accurately every detail of the physical structure of objects viewed or to listen to every word of verbal communication. The mind takes the trouble to perceive only as much as is necessary to classify the object of the message; the rest is ignored as redundant. In other words, things are perceived as wholes using a minimum of clues or symbolism. Advertising for many familiar brands is instantly recognizable through the use of imagery and symbols—the prominent Oxo logo and red pack, Hovis logo, the aqua and black Heinz beans tin.

This is also of considerable significance in respect of much packaging research, showing that a great deal of printed material—ingredients lists, for instance—

is simply ignored. Shoppers reach for a familiar or interesting *total* configuration of elements. This helps to explain the persistence of purchasing according to habit.

Much advertising thinking for many years was based on an essentially one-way flow of communication. This was the source of much of the militaristic conversion rhetoric: that consumers could be persuaded at will. However, we know that individuals selectively expose themselves to media, selectively distort the content of the messages, and selectively retain the information (Rowan, 1968). This is crucial to understanding how advertising works. Individuals are rarely information-seeking mechanisms, but more often information-scanning—selecting the bits that fit, and rejecting the rest.

Here the central metaphor is a complex set of filter systems by which individuals block out unwanted or unneeded signals but let others pass by. The question then becomes, what are the influences or factors that activate the filter systems? How conscious or unconscious are they?

Practitioners in the business of advertising and the use of persuasive communications, as well as the public, are used to thinking in terms of 'what does advertising do *to* people?'. I have suggested that in its strictest sense persuasive communications do not do anything to people unless they are *allowed* to—an 'active voice' perspective. Consequently, a more fruitful way of looking at the manner in which advertising works is to turn the question around. Rather than asking what advertising does to people, the question should more properly be: 'What do people do *with* advertising?'

Much of the most illuminating and rewarding thinking on the subject of mass communications in society, whether this is advertising or persuasive communications of other sorts, comes from social anthropologists. Indeed, there has been a trend in Britain in recent years to bring the social sciences together and look at them holistically as a multidisciplinary approach to understanding culture. Phenomenology, semiology and structuralism add significantly to the richer understanding of consumption behaviour that classic economic models provide. Mary Douglas (1982a) and many others (e.g. Nicod, 1980) have contributed significantly to the growing literature on food as communication: the ritual significance of different food patterns, the importance of structure, the stability of the structure of the British meal, the layers and levels of meaning that different foods have in relation to different occasions, different members of the family, non-family members and so forth.

It is in this area that the most fruitful sources of insight for the development of packaging and advertising for food products lie.

BRANDS VERSUS PRODUCTS

A basic finding in product research is that the ability to discriminate between tastes (although not textures) is very blunt indeed. This is demonstrated through either blind taste tests of different foods or, more interestingly, tests of the same foods in different guises. For example, a sample of 100 housewives showed a significant preference for a piece of meat labelled English lamb compared to a piece of meat labelled New Zealand lamb when, in fact, both pieces were cut from the same New Zealand joint. Changing the packaging can change the taste.

How then is it that people choose one brand rather than another? The phenomenon is more pronounced in drink than it is with food. Nevertheless, it is worth asking why certain brands of food are bought habitually, and why certain brands obtain high rankings in research asking people to rate them on dimensions such as 'good-quality ingredients', 'used by discerning hostesses' and so forth, when blind product tests show little difference and price is not a factor?

A partial answer is in the distinction between a product and a brand—between what the manufacturer makes and what the consumer buys and finds satisfying. The product is merely the physical stuff, the ingredients laid out on a table or packaged in a plain brown bag. The brand is the physical product combined with packaging (shape, materials and design), advertising and other promotional activities. Thus, what makes a brand is the combination of the two which involves symbols, images and feelings. This has been described (Cooper and Pawle, 1983) as a sort of attachment or 'symbiosis' which consumers have for 'their' brands and the advertising that surrounds them.

After Eight Mints could have been called Rowntree Peppermint Creams, a more or less accurate functional description of what is in the box. However, the dark brown envelopes, the name, the green and gold clock motif and the archetypal advertising theme add many layers of meaning that in the early stages of the brand's life satisfied the need guests felt to present their hostess with a gift—a culturally determined ritual. Mr Kipling Cakes, Kellogg's Corn Flakes, Ovaltine, Black Magic are other examples of brands of food products where product, packaging and advertising combine to reinforce choice and add to the pleasure of consumption because the total presentation carries meaning and symbolism that have a personal or social *utility* for the consumer.

RATIONAL, SENSUAL AND EMOTIONAL APPEALS IN FOOD

It is useful, when considering how to present a food product, to think in terms of three different sorts of appeal: appeal to the reason, appeal to the senses and appeal to the emotions. Our attachment to anything we buy is a combination of these three appeals: however, it is obvious that the relative importance of these appeals will vary according to the nature of the food and its consumption occasion, combined with its competitive position in the market.

Rational Appeals: Problems with Nutrition as an Appeal

Although, increasingly, nutritional information is offered as supporting evidence, it is conspicuous that very few brands of food are advertised solely on a nutritional basis. At least part of the explanation is that women by and large do not choose brands of foods solely on the basis of their nutritional value. Many other criteria come into play according to the meal, e.g. speed of preparation, variety, price.

The main meal of the day is the one in which nutrition counts most. With other meals, such as a snack meal for children after school, something fun and interesting takes priority. With the midday meal, if the housewife is by herself, the priority is that the food be cheap and quick and easy to prepare. This is a blunt and simplified scheme. Nicod and Douglas (1974) list more complex structures within which food is chosen and consumed in British culture. Nevertheless, it illustrates the main point—that nutritional value *per se* is not always of particularly high priority.

Despite the increased interest in nutrition, data from the massive British Nutrition Foundation Study reported in 1984 suggests that the link between attitudes and behaviour is still rather weak. To quote from the report: 'A recurrent theme throughout our research has been the discrepancy between attitudes and behaviour, between what they say they believe about food and what they actually eat.'

So what up-to-date attitude surveys indicate—e.g. the survey by Birds Eye (1983) (see Table 13.1)—is that naming nutritional value in this context is a little like voting for motherhood and against sin.

Figure 13.1 How important are each of the following when selecting the food you eat.

Level of importance:	Very	Quite	Not very	Not at all	Don't know
Taste	83%	14%	1%	0%	0%
Value for money	67%	26%	5%	1%	1%
Nutritional value	62%	27%	9%	2%	2%
Price	55%	33%	10%	2%	1%
Family influence/ preference	52%	30%	10%	5%	3%
Convenience	24%	41%	29%	5%	0%

Source: Bird's Eye Report (1983).

That 62 per cent named nutritional value as 'very important' in selecting the food they eat is in part a socially normative rationalization which does not necessarily relate to behaviour.

A further analysis of the British Nutrition Foundation data linked with National Food Survey data underlines the unreliability of what people claim about their behaviour (see Table 13.2). The same point is made by looking at foods people claim to eat *more* of (see Table 13.3). Which is not to deny that segments of the population *are* buying healthier foods—sales of yoghurts, fruit juices, wholemeal breads, high-fibre foods and others are increasing. However, the expressed positive attitudes towards these foods is far higher than sales reflect: thus the *gap* between expressed attitudes and behaviour is still considerable.

Figure 13.2 Consumers' claimed decrease in consumption.

	BNF % claiming to have given up or cut back over the last year or so 1982	NFS % changes in consumption 1980-1982
Cakes	34	−1
Sugar	30	−8
White bread	25	−1
Biscuits	25	+5

Source: BNF (1984): *Eating in the Early 1980s.*

Figure 13.3 Consumers' claimed increase in consumption.

	BNF % claiming to eat more over last 2 years 1982	NFS % changes in consumption 1980-1982
Brown/wholemeal bread	21	−3
Fruit	19	−9
Cheese	14	−2
Eggs	12	−5

Source: BNF (1984): *Eating in the Early 1980s.*

The British Nutrition Foundation report gives further insight into the problems involved in narrowing this gap by comparing authoritative sources with action. Table 13.4 shows that asking consumers which they believe, tells you far more about the status of various sources than about the extent to which people will act on information from these sources.

More intriguingly, communication about nutritional value can unwittingly reinforce a prejudice against a food.

The history of soya products in this country has been, from many standpoints, a rather disappointing one. Nutritionists are sorry to see such a cheap source of protein ignored and manufacturers have been surprised to find such a relative lack of acceptance of products which, on the face of it, should offer considerable advantages. There are two points to be made in this respect. First, product tests have demonstrated that the products simply are not good enough, a finding that might have been predicted from an anthropological analysis. Meat,

Table 13.4 Sources of information believed and acted on.

	Adults		Children (11-15)	
	Would believe %	Ever acted on %	Would believe %	Ever acted on %
Doctors	90	31	93	10
Dieticians	72	8	65	3
Nutritionists	70	3	64	1
Government information	50	2	36	1
Societies	46	4	53	—
Cookery books	43	2	42	1
Parents	38	6	61	27
Schools	29	2	27	5
Magazines	18	3	18	1
TV	17	2	19	5

Source: BNF (1984): *Eating in the Early 1980s*

a central part of the structured aspect of the British diet, is a food regarding which people are highly discriminating and the possibility of rejection on taste grounds is very high.

However, a subtler point is the way soya products were originally presented through advertising. It is an extraordinary technical achievement to produce 'meat from plants', and in the early stages of these products, some advertising and a great deal of editorial concentrated on this, on the assumption that the consumer would find this as interesting as the manufacturer. But, of course, this overlooks the crucial importance of taste. Consumers wish to be reassured that the product both tastes good and fits into a normal meal occasion. The last thing they want to know or need to know is that it is abnormal in origin, never mind how nutritious and cheap.

Successful uses of Nutritional Claims

Obviously, nutritional claims *can* be made very advantageously—to return to the earlier metaphor—when the filters open easily and messages are readily accepted. In the current climate they form at least part of the total advertising communications. As the case of the Bread Advertising Group illustrates, the nutritional claims are most acceptable when the bread is both familiar and (most importantly) widely enjoyed in taste terms. In 1978 the Bread Advertising Group decided to commit money to advertising generic white bread. The advertising task was to develop a campaign to help increase consumption of all white bread; white bread consumption had been steadily decreasing for many years.

The first task was to establish why this was the case. To oversimplify greatly, the reasons were several: competition from many other foods, particularly for snack meals; price; and, most surprisingly, something that could be called status. Table 13.5 shows how women rate bread as a source of high-status food values (protein) and as a source of low-status food values (carbohydrates).

Figure 13.5 Housewives' beliefs about food.

	Sources of protein %		Sources of carbohydrates %
Meat	49	Bread	48
Cheese	33	Potatoes	34
Eggs	29	Cakes	26
Fish	24	Sugar	18
Milk	16	Butter	15
Vegetables	7	Meat	14
Bread	5	Milk	14

Source: Study conducted for Bird's Eye by Research Bureau Ltd (1978).

The advertising task was to determine the most effective way of conveying facts about the food values of bread. The first problem the exploratory research addressed was: why is white bread so undervalued, and is there an underlying resistance to accepting facts about nutritional values of bread? The major findings were that most women claim to like, even 'love', bread. At the same time they feel guilty eating it because they believe it contains only empty calories: bad values (fattening) or none at all rather than good values (protein source). But underlying all of this were residual beliefs in bread's goodness, stemming from the meaning of the phrase 'staff of life', the communal breaking of bread rituals, and the role of bread remembered in their own childhood—symbolically linked with maternal warmth and security.

Experiments were conducted with a number of different ways of conveying the nutritional facts about the food values of bread. Serious and ostensibly authoritative presentations of food facts were either misinterpreted or misunderstood; strictly emotional appeals describing bread as 'the staff of life' were perceived as evocative but insubstantial. The most convincing presentation was a comparative claim using what women already knew about foods containing protein as a frame of reference: hence the claim 'penny for penny, bread is a cheaper source of protein than milk, eggs, cheese' used in the television commercials.

The launch commercial embodied a number of elements that contributed to the overall impression of warmth and seriousness. As a piece of communication this comparison was the most meaningful way of increasing awareness of the protein value of bread, and the advertisement used the context of bringing up a family and the phrase 'the family's greatest supporter' to further underline the ubiquity of bread. It is the balance of emotional and factual information that makes the communication meaningful.

The success of St Ivel Gold is another example. The brand is a spread containing half the fat of margarine. A great deal of advertising research on various ways of presenting this claim concluded that the claim *on its own* communicated very much *less* effectively than when it was combined with a more balanced theme which showed visual clues to normal healthy living and emphasized the fresh 'buttery' taste—through the soundtrack linked with visual devices of an appetising butter sculpture.

Emotional Appeals: The Importance of Appropriate Imagery

The word 'imagery' often carries negative connotations, suggesting tricks employed by advertisers—dishonest illusion rather than honest fact. What is meant by food imagery are the elements beyond factual nutritional components, or descriptions of convenience benefits, that influence how food is chosen and used. There are basically three areas where foods require carefully designed imagery to link comfortably with existing attitudes and beliefs.

Links with traditional occasions and rituals

As I have mentioned, all societies have culturally and historically defined eating patterns. The ritual of Sunday lunch, and breakfast and tea patterns, are obvious examples in the UK. Nicod (1980) distinguishes between two aspects. On one hand are the highly structured parts of the British food system which centres on the main meals in which innovations are mainly of better quality in the more traditional foods and more variety in the trimmings. Oxo over many years has linked itself firmly to the central tradition of the Sunday joint. On the other hand, brands such as Horlicks, which for many years was closely allied to a stable *social* pattern but unstructured *food* pattern—the ritual of marking the closing of the day with family consumption of hot bedtime drinks—has to redefine itself in the light of the decline in this habit.

One of the prominent rituals in any society is the giving of presents. Advertising and packaging play a considerable role in the choice of foods as gifts. Giving is one of the most subtle social transactions and the appropriateness or inappropriateness of certain foods in certain circumstances is culturally prescribed. In this culture, chocolate assortments are acceptable expressions of a range of messages from giver to receiver. For example, Black Magic has clear and long-standing romantic associations that would be clearly inappropriate and embarrassingly intimate in certain circumstances. By the same token, Dairy Box has another set of associations attached to it: thanks to the baby sitter, the neighbours, the more distant (psychologically) relatives, friends. What is interesting in respect of gifts is the crucial importance of both members of the transaction understanding exactly what is being symbolized by the gift. Unlike certain other categories where a certain ambiguity is useful in extending the appeal, with gifts, ambiguity can be a drawback.

It is ironic that a category where there are discernible product differences in terms of choice of covering, choice of filling, number and type of chocolates per box and type of chocolate, advertising has almost no scope for emphasising these differences since the stressing of the physical features suggests a value for money orientation entirely inappropriate in a gift.

In sharp contrast to chocolate assortments are the individual bars of chocolate or the newer category of muesli-based 'health' bars purchased for personal consumption. Here the requirement is exactly the opposite. The advertising must show very clearly exactly what is in the bar: plain or milk chocolate, wafers, biscuit, nuts, sultanas, nougat, natural sugar, etc. People who buy and eat chocolate bars themselves have very clear ideas of what they like and scan the advertising for exactly what the bar contains. What consumers are looking for is a particular combination of flavours, ingredients and textures, hence the considerable attention paid to illustrating and demonstrating exactly what is on offer.

The importance of user imagery

The second area where imagery is important is user imagery. What kinds of people use this brand and how can imagery be constructed to overcome prejudices? Here advertising must be extremely sensitive to timing and responsive to changing public beliefs.

The development by British Bakeries of the advertising for the Windmill Bakery range of brown bread in 1980 is a good example of constructing both user and occasion imagery to encourage predominantly white bread buyers to try the brand, using advertising especially designed to overcome some of the resistances held by users of white bread towards the regular use of brown bread. The aim was to use the advertising to make the bread more emotionally accessible and thereby increase likelihood of trial.

The imagery of 'health' food products has been discussed by many writers (e.g. Eyton, 1982) who provide considerable insight into the difficulties of foods identified as 'health' foods (as opposed to 'healthy' foods) in gaining *mass* acceptance. Symbolically, these foods carry rebellious meanings—of unorthodox lifestyles and alternative value systems. This is changing gradually, but because food and eating habits are rooted in the social order, resistances and prejudices can be very powerful so that change on a large scale is inevitably very slow.

Qualitative research amongst users of both kinds of bread explored this hypothesis and examined the range of other, more consciously accessible factors that inhibited most women from buying brown bread *regularly* and a large proportion of women from buying brown bread at all. Most interestingly, the research suggested that amongst users of both kinds of bread, levels of expressed interest in brown bread were considerably higher than actual consumption. Most occasional users of brown bread and some of white bread described brown bread as wholesome, natural, good for you. It was described as natural, uncontaminated by artificial preservatives. The 'status' of brown bread was generally seen to be high.

However, despite the high status or desirability of brown bread, this research identified a number of resistances to actually buying it and using it regularly.

Children were described as preferring white bread because it is 'smooth', 'white', 'pure' and 'clean'. Conversely, brown bread is believed by children to be 'boring', 'dry', 'crumbly', 'dirty', with some children reported as 'loathing' brown bread. However, these opinions attributed to children may well be projections of mothers' own feelings, or at least excuses or scapegoats to avoid using brown bread.

Concealed behind these overt attitudes are women's wishes to avoid imposing brown bread on children for fear of being rejected themselves as 'unpleasant' mums. Extreme images of 'women who only served brown bread' were described as 'no fun'. Furthermore, since children often prepare bread for themselves, mothers feel a necessity to 'give in' to children on this point if they wish them to eat bread at all.

The beliefs attached to brown bread suggest that, to women who eat only white bread, brown is a 'complex' subject with lots of brown breads to choose from. White bread is 'straightforward'. Along with beliefs such as that brown bread is not suitable for making sandwiches (it is felt that it is crumbly, goes hard and dry, etc.), other defences come into play such as not being able to make bread and butter pudding, that 'other people' do not like it.

Perhaps most important, white bread is 'normal'; it is something you depend on and are used to as part of the daily routine. It is simply 'there' to be used; it does not interfere as brown bread is perceived to do. Extreme projections of 'women who use only brown bread' seen through the eyes of user of white bread were most revealing. 'Health food addicts', 'cranky', 'faddy', 'obsessed with their bowels', 'odd in some ways', were the terms that women used, illustrating the psychological distance from 'normal' women with 'ordinary families' (in anthropological terms, a rejection of a symbol of unorthodox lifestyles and values). It appeared that the cultural history of brown bread as *virtuous* was contributing to the social distancing of it from the habits of ordinary women.

Thus, the major conclusions stemming from this research led to an advertising campaign that positioned a new brown bread as 'normal': a bread that is used by ordinary women with growing families; that suggests familiar healthy outdoor energies; and that suggests a wider range of 'normal' uses. The launch commercial uses a windmill device and ties it to the idea of people working up a healthy appetite with the motion of the sails compared to and intercut with people 'windmilling' their arms. The activity can be part of their job—farmers scaring crows, people chopping logs, policemen directing traffic. Or part of their recreation—fast bowlers warming up, gymnasts. Or just *joi de vivre*—children swinging their arms to keep balance, or turning cartwheels.

Brown bread is gradually moving from being a 'health' food to being a 'healthy' food. Once it has been adopted by large numbers of people and is thereby perceived as normal, *then* advertising may usefully focus on its nutritional aspects—e.g. Allinson's: bread with 'nowt' taken out. Other foods for which this social normalizing strategy could be employed are lentils or brown rice, or indeed any 'health' food that can be shown to have an acceptable flavour in blind test.

Experience with bran cereals illustrates subtle differences in advertising messages. The aim of a manufacturer, such as Kellogg's, marketing a number of different brands is, of course, to appeal to as many different types of people as possible so that those brands are not competing with each other. Consequently, All-Bran is at one end of the 'serious' fibre spectrum while Bran Flakes and Sultana Bran are towards the other.

The advertising for both Allinson's and Kellogg's has evolved in response to increased public interest in fibre, interest fuelled by editorial, television coverage and, most recently, by the success of *The F-Plan Diet* by Audrey Eyton (1982). The history of All-Bran illustrates the crucial importance of timing.

From the 1950s the communications about the brand were highly factual, dominated by laxative claims. The brand was used both by users of constipation remedies and people who feared constipation—a very small market indeed. By non-users, the brand was described as medicinal and clinical, and its users as peculiar or 'odd' with an unhealthy preoccupation with their insides: in a word, thoroughly abnormal ('not me') types of people. An advertisement of the early 1970s comparing the fibre in All-Bran to that in cabbage communicated successfully with users aware of the functional virtues of cabbage, but non-users thought only of the taste of cabbage and the associations were almost entirely negative: school dinners, ugly smells, a watery, tasteless result. So that the message *received* was 'cabbage is boring and so is All-Bran'. A campaign that ran in the early 1980s with the headline 'Eat yourself fit' recognized the need to emphasize generally healthy living rather than to draw attention to a problem.

Bran Flakes, on the other hand, has a potentially wide appeal on *taste* grounds and the advertising, using the phrase 'very, very tasty', addresses itself to the prejudices of ordinary breakfast cereal eaters—that bran-based products, while obviously virtuous and good for health, may not taste as delicious. In this case the advertising acts as reassurance.

Source credibility

Finally, there is the issue of source authority. And here there are several aspects worth commenting on; for instance, the problem of introducing foreign flavours into the British diet, and the need for reassurance in the advertising of manufactured foods in general.

There is no doubt that an important trend in food purchasing is an interest in foreign foods and that a major role for advertising foods from other countries to a *mass* audience is to recognize the need for what might best be described as 'safe exoticism' and to select appropriate imagery. This is often taken to mean that the foods themselves must be formulated to make the flavours more bland, less sharp and distinctive, because of a notion that English tastes are 'conservative'—often taken to mean bland. This seems to be a serious misunderstanding of English tastes, which simple observation suggests are far from bland: English mustard, Worcester sauce, the strong vinegar base in many condiments, sharp Cheddars, strongly flavoured beers. What English consumers may require in certain product categories is not bland flavours but familiar (not foreign) packaging and presentation. Both Birds Eye and Findus understand particularly well the importance of a balance between exoticism and familiarity in the design of food packaging intended for use by the whole family. This reassurance is directed mainly to women and shows an understanding of the problems of housewives wishing to introduce what may be interesting and varied foods (satisfying her own needs) to conservative husbands and children.

On the other hand, shorthand stereotypes of countries are often used successfully, and increasingly, direct imports are gaining popularity—Chinese, Indian and Mexican are the most conspicuous examples. Many years ago, Ski brand yoghurt was a very much more specialized food that it is today and benefited from the (believed) Swiss concern for health. Subsequently, Alpen used a similar theme for its muesli-type cereal. The use by Wall's ice-cream (Cornetto) of the stereotype of Italian expertise in ice-cream is another example.

Anthropological analysis would suggest that the introduction of foreign elements would be resisted in the *central* part of the British meal, the family dinner. However, unusual, foreign or new foods can be acceptable on a mass scale in children's food or snacks or light meal occasions. Indeed, following this logic, if meal patterns fragment further with more and more small snack meals replacing the traditional and ritualistic family occasions, many of the traditional resistances towards newness and foreignness will begin to disappear.

Although manufactured food is widely used and widely acceptable, it is a generalization that obscures many of the more subtle issues in packaging and advertizing. And while it is an overstatement to say that women view all prepared, manufactured food with suspicion, there are a number of both expectations and concerns that must be recognized.

Here is is worth distinguishing the role of the packaging from the role of the advertising. The mere presence of a packet, tin, sachet, bag indicates that the food has been manufactured. If it has been manufactured, it is because the preparation of a familiar but time-consuming food is made quicker or easier—packet sauces, for instance. Or it is because it is something that women are unable or unwilling to make themselves—flaky pastry, for instance. In all of these cases, it is quite unnecessary to indicate on the packet or in the advertising that the product is fast or convenient: the packaging on its own conveys this message. The role for the advertising is to reassure purchasers that the product tastes as good as the 'real thing'. On the other hand, if the product is one of the numerous sorts of foreign foods that are gaining widespread acceptance, such as cannelloni, lasagne, pizza, *coq au vin,* the role for the advertising is two-fold: to indicate *where* in the total meal pattern this new food fits, and to provide reassurance that it is widely acceptable (by other people, or by the family) on taste grounds.

Reassurance can be conveyed in many ways. With well-known and respected companies such as Heinz, Kellogg's, Cadbury, St Ivel, often the name alone is a guarantee. The mechanism is the same sort of mechanism that operates in any other field. Companies which have manufactured products for many years, which have become household names, which conduct thorough product testing to keep their standards high, and which constantly improve and innovate, have acquired reputations for a standard that acts as powerful assurance. 'If it's made by Kellogg's/Heinz, etc., it is likely to meet some sort of minimum standards', goes the train of thought. It does not necessarily guarantee a personal subjective preference for size, shape, flavour, colour, but it will act as a recommendation for the first trial at least.

Presenters assumed to have food or cooking expertise are also used to lend authority. However, a more discerning public is increasingly likely to suspect mercenary motives and rely more and more on the reputation built up by the manufacturer.

Retailers are increasingly taking over the traditional role of the manufacturer in the guarantee of quality. Marks & Spencer foods enjoy an excellent reputation and their prepared foods are ones that consumers are prepared to pay a premium for. Sainsbury, Waitrose, Tesco are all investing in good-quality, highly acceptable foods providing serious competition for the established food manufacturers.

I have commented on the message that packaging construction carries: sachets, tins, frozen boxes are by definition convenient. The generic claim for the product's existence is inherent in how it is packaged. However, what does the packaging do to influence brand preferences in a market where there are several brands of tinned, boxed, frozen, sachet-ed foods? Here the significance of package design can play an important part in influencing preference.

A notable example is Mr Kipling Cakes. The most important feature in the success of this range of cakes is that they are of excellent quality. However, the packaging and the advertising were deliberately designed to convey this quality by the choice of a hessian-style background, the use of an old-fashioned typeface (arguably difficult to read), the choice of angle to photograph the cake—opened and appetising, looking irregular and home-made. The choice of a name, Mr Kipling, symbolised the idea of a bespoke master baker and the carrying through of this theme in the advertising created a personality and style for the brand that has virtually given it a life of its own. At a rational level, Mr Kipling Cakes are known to be manufactured by a large, anonymous firm. However, the totality of the excellent quality product, old-fashioned packaging and consistently personalized advertising support the symbolic wish for providing home-made cakes. In this sense, the presentation acts to reduce guilt at providing a food that it is traditional to make oneself, and thereby serves to both reassure and rationalize choice.

CONCLUSION

It is clear that food patterns in Britain are being influenced by many, many factors. Most significant are the desire of more affluent, sophisticated, widely travelled (actually or vicariously) people to experiment with new types of food, the increasing demand for better quality ingredients and flavours, the *interest* in natural, healthy foods (even if the behaviour lags behind). Furthermore, social and demographic trends—more entertaining, less formal, ritualized meal times, more eating out, more women working—affect food choice.

Advertising and packaging obviously play a part. But their contribution will only be successful when the manufacturer is in tune with what people want and is sensitive to the subtleties of the many different roles that food plays in people's lives.

REFERENCES

Atkinson, Paul (1980) 'The symbolic significance of health foods'. In *Nutrition and Lifestyles,* ed. M. R. Turner, Applied Science Publishers, London.

British Nutrition Foundation (1984) *Eating in the Early 1980s.* Market Research Bureau International Ltd, London.

Cooper, Peter and Pawle, J. (1983) 'Brand personality—its place in the strategy planning process', *Proceedings of the British Pharmaceutical, Marketing and Research Group Conference 1982,* Bristol.

Douglas, Mary (1982a) 'Food as a means of communciation'. In *In The Active Voice.* Routledge & Kegan Paul, London.

Douglas, Mary (1982b) 'Goods as a means of communication'. In *In The Active Voice.* Routledge & Kegan Paul, London.

Douglas, M. and Nicod, M. (1974) 'Taking the Biscuit: the Structure of British Meals', *New Society*, **19**, Vol. 30 No. 637.

Eyton, Audrey (1982) *The F-Plan Diet.* Penguin, London.

Festinger, Leon (1957)*A Theory of Cognitive Dissonance.* Harper & Row, New York.

Koffka, K. (1935), *Principles of Gestalt Psychology.* Harcourt Brace, New York.

Nicod, M. (1980) 'Gastronomically speaking—food studied as a medium of communication'. In *Nutrition and Lifestyles,* ed. M. R. Turner, Applied Science Publishers, London.

Rowan, John (1968) Selective Perception and Advertising Research, Part 2. *Admap,* **4**(8), 358-365.

Index